Stochastic Simulation

Stochastic Simulation

BRIAN D. RIPLEY

Professor of Statistics
University of Strathclyde
Glasgow, Scotland

WILEY-
INTERSCIENCE

A JOHN WILEY & SONS, INC., PUBLICATION

Library of Congress Cataloging-in-Publication is available.

ISBN-13 978-0-470-00960-4
ISBN-10 0-470-00960-8

10 9 8 7 6 5 4 3 2 1

Preface

This book is intended for statisticians, operations researchers, and all those who use simulation in their work and need a comprehensive guide to the current state of knowledge about simulation methods. Stochastic simulation has developed rapidly in the last decade, and much of the folklore about the subject is outdated or fallacious. This is indeed a subject in which "a little knowledge is a dangerous thing!" Although this *is* a comprehensive guide, most of the chapters contain explicit recommendations of methods and algorithms. (To encourage their use, Appendix B contains a selection of computer programs.) Thus, this book can also serve as an introduction, and no prior knowledge of the subject is assumed.

Simulation is one of the easiest things one can do with a stochastic model, which may help to explain its popularity. Although easy to perform, some of the "tricks" used are subtle, and the analysis of what has been done can be much more complicated than is apparent at first sight. Simulation is best regarded as mathematical experimentation, and needs all the care and planning that are regarded as a normal part of training in experimental sciences. The general mathematical level of this book is elementary, involving no more than a first course in probability and statistics. A notable exception is those parts of Chapter 2 that deal with the theoretical behavior of random-number generators, which contain a number of applications of number theory. All the necessary mathematics is developed there, but some prior knowledge of pure mathematics will help a great deal. Random-number generators are so fundamental that the reader should eventually tackle Chapter 2 unless he or she is *sure* that all the generators he or she uses are adequate (that is, have been checked by someone who understands that chapter). It might be disastrous to believe in your computer manufacturer!

Chapters 3 and 4 cover drawing realizations from standard probability distributions and stochastic processes. The emphasis is on methods that are easy to program (compact and with a simple logic, therefore easy to check). These are particularly suitable for personal computers. A small number of workers have specialized in developing faster and increasingly more complex algorithms. These are referenced but, in general, not described in detail. The coverage of *methods* was comprehensive at the time of writing.

Even statisticians often fail to treat simulations seriously as experiments. Even more is possible in the way of design since the randomness was introduced by the experimenter and hence is under his or her complete control. Such techniques are described in Chapter 5 under the heading of "variance reduction." A general knowledge of the statistical design of experiments is helpful here and essential to a competent practitioner of simulation. The analysis of the output of many simulation experiments, for example queueing systems, is also more complicated than many users suppose, although not as difficult as the literature makes out! This topic is discussed in Chapter 6.

Chapter 7 discusses many novel uses of simulation. It can be used, for example, in optimizing designs of integrated circuits and in fundamentally new ideas in statistical inference.

The literature on simulation is vast, and I have made no attempt to cite comprehensively. There are several published bibliographies, but a lot of the work has been superseded or is misleading.

The exercises vary considerably in difficulty. Some are routine exercises in developing algorithms from general theory or in providing illustrative examples. Others are of an open-ended nature; they suggest experiments to be done and demand access to a computer (although the humblest personal computer would suffice).

Simulation has long been a cinderella subject, particularly in statistics. I hope this book shows that it raises fascinating mathematical and statistical problems that demand attention.

BRIAN D. RIPLEY

Glasgow
October 1986

Acknowledgments

I am indebted to everyone who has taught me about simulation or has been prepared to share their experiences with me, in particular, Anthony Atkinson and Luc Devroye. The manuscript was typed with great efficiency by Lynne Westwood. The figures were produced on equipment funded by the Science and Engineering Research Council.

B.D.R.

Contents

1 Aims of Simulation   **1**

1.1 The Tools, 2
1.2 Models, 2
1.3 Simulation as Experimentation, 4
1.4 Simulation in Inference, 4
1.5 Examples, 5
1.6 Literature, 12
1.7 Convention, 12
    Exercises, 13

2 Pseudo-Random Numbers   **14**

2.1 History and Philosophy, 14
2.2 Congruential Generators, 20
2.3 Shift-Register Generators, 26
2.4 Lattice Structure, 33
2.5 Shuffling and Testing, 42
2.6 Conclusions, 45
2.7 Proofs, 46
    Exercises, 50

3 Random Variables   **53**

3.1 Simple Examples, 54
3.2 General Principles, 59
3.3 Discrete Distributions, 71
3.4 Continuous Distributions, 81
3.5 Recommendations, 91
    Exercises, 92

ix

4 Stochastic Models **96**

4.1 Order Statistics, 96

4.2 Multivariate Distributions, 98

4.3 Poisson Processes and Lifetimes, 100

4.4 Markov Processes, 104

4.5 Gaussian Processes, 105

4.6 Point Processes, 110

4.7 Metropolis' Method and Random Fields, 113

Exercises, 116

5 Variance Reduction **118**

5.1 Monte-Carlo Integration, 119

5.2 Importance Sampling, 122

5.3 Control and Antithetic Variates, 123

5.4 Conditioning, 134

5.5 Experimental Design, 137

Exercises, 139

6 Output Analysis **142**

6.1 The Initial Transient, 146

6.2 Batching, 150

6.3 Time-Series Methods, 155

6.4 Regenerative Simulation, 157

6.5 A Case Study, 161

Exercises, 169

7 Uses of Simulation **170**

7.1 Statistical Inference, 171

7.2 Stochastic Methods in Optimization, 178

7.3 Systems of Linear Equations, 186

7.4 Quasi-Monte-Carlo Integration, 189

7.5 Sharpening Buffon's Needle, 193

Exercises, 198

References **200**

Appendix A. Computer Systems   **215**

Appendix B. Computer Programs   **217**

  B.1 Form $a \times b$ mod c, 217
  B.2 Check Primitive Roots, 219
  B.3 Lattice Constants for Congruential Generators, 220
  B.4 Test GFSR Generators, 227
  B.5 Normal Variates, 228
  B.6 Exponential Variates, 230
  B.7 Gamma Variates, 230
  B.8 Discrete Distributions, 231

Index   **235**

Stochastic Simulation

CHAPTER 1

Aims of Simulation

The terminology of our subject can be confusing, with some authors insisting on shades of meaning that do not have widespread agreement. A dictionary definition of "to simulate" is

> Feign, ..., pretend to be, act like, resemble, wear the guise of, mimic, ... imitate conditions of (situation etc.) with model, for convenience or training ....
>
> *Concise Oxford Dictionary*, 1976 ed.

In everyday usage "simulated" has a derogatory ring, but the value of simulators in training pilots is also recognized. In its technical sense simulation involves using a model to produce results, rather than experiment with the real system under study (which may not yet exist). For example, simulation is used to the explore the extraction of oil from an oil reserve. If the model has a stochastic element, we have *stochastic simulation*, the subject of this monograph.

Another term, the *Monte-Carlo method*, arose during World War II for stochastic simulations of models of atomic collisions (branching processes). Sometimes it is used synonymously with stochastic simulation, but sometimes it carries a more specialized meaning of "doing something clever and stochastic with simulation." This may involve simulating a different system from that under study, perhaps even using a stochastic model for a deterministic system (as in Monte-Carlo integration). We will not use Monte Carlo except in the conventional terms "Monte-Carlo integration" and "Monte-Carlo test."

Simulation can have many aims, which makes it impossible to give universal guidelines to good practice. Tocher (1963) wrote one of the first texts on the subject. His title was *The Art of Simulation*, and simulation is still an art despite a much greater understanding of the simulator's toolkit. The aim of this volume is to display those tools in their most useful form with guidance about their use.

1

1.1. THE TOOLS

The first thing needed for a stochastic simulation is a source of randomness. This is often taken for granted but is of fundamental importance. Regrettably many of the so-called random functions supplied with the most widespread computers are far from random, and many simulation studies have been invalidated as a consequence.

Digital computers cannot easily be interfaced to a truly random phenomenon such as the electronic noise in a diode. All random functions in common use are in fact pseudo-random, which is to say that they are deterministic, but mimic the properties of a sequence of independent uniformly distributed random variables. Their essence is unpredictability. Consider for example the following sequence

$$13, 8, 1, 2, 11, 14, 7, 12, 13, 12, 17, 2, 11, 10, 3, \ldots$$

It is generated by a simple deterministic rule, but no one had guessed what the rule was or what the next number is at the time of writing. (Exercise 1.1 will give the game away, but try to guess first.) The algorithms commonly used are similar, and much mathematical analysis has gone into the question of how well they do mimic a random sequence.

Only occasionally does one want independent, uniformly distributed random variables. However, they are a useful source of randomness that can be turned into anything else. Chapters 3 and 4 consider tools to make samples of all the standard distributions and stochastic processes from this source of randomness.

Simulation for us is about sampling from stochastic models. Too much emphasis has been placed in the literature on producing the samples and too little on what is done with those samples. Any stochastic simulation involves observing a random phenomenon and so is a statistical experiment. Statisticians, even experts in the design of experiments, are notoriously bad at designing their own experiments! There is even more scope for designing a simulation experiment than a real one, for the randomness and the model are under our complete control. Thus techniques for the design and analysis of simulation experiments are important tools and still an under-researched area.

1.2. MODELS

A stochastic simulation is of a *model*, and the aims of simulation are closely connected to those of modeling. So, why model? Within the scope of statistics and operations research we can usefully identify two principal reasons:

1. *To summarize data.* A very common example is the general linear model of statistics as used in regression and the analysis of variance.
2. *To predict observations.* A regression equation can be used to predict a response under new conditions or to find a combination of control variables giving an optimum response. This "what if" use of models is the basis of much of operations research.

It is also useful to consider two classes of a model. Models can either be *mechanistic* or *convenient.* For example, the general linear model is merely convenient whereas the models of genetics are thought to represent the actual mechanisms. The models of the physical world used by engineers are usually both deterministic and mechanistic, whereas most stochastic models are convenient. Either type of model can be used to help understand, to predict, or to aid decision-making. An example of the latter is the "convenient" models of errors in agricultural field trials which are used to help disentangle the true differences in fertility of plant varieties from the fertilities of the plots in which they were grown.

To make use of a model one has two choices:

1. To bring mathematical analysis to bear to try to understand the model's behavior. This is very easy for a general linear model but nigh impossible for a complex queueing system or for the equations of fluid flow in a complex structure such as a rock. The work involved is usually laborious (although if one is lucky it may already have been done). There are also likely to be necessary approximations and questionable assumptions.
2. To experiment with the model. For a stochastic model the response will vary, and we will want to create a number of *realizations* (sets of artificial data) for each set of parameters.

Sometimes one of these choices may be unfruitful. We might not be able to make progress by analytical means or might not have the resources to simulate the model. (It is almost always possible to simulate a well-defined model given sufficient resources.)

The choice of analysis or simulation will depend on the purpose of modeling. Simulation is good at answering specific "what if" questions whereas analysis almost always deepens understanding of the model. One neglected use of simulation is a hybrid approach: do a simulation experiment, analyze it to produce a "convenient" model, and use *this* model for predictions and decisions.

The cost analysis is rapidly tilting in favor of simulation as computer time becomes ever cheaper and mathematicians remain scarce. It may be incredible to younger readers that Cox and Smith (1961) reported a simulation performed with the aid of a slide rule (a mechanical device to perform multiplications and evaluate standard functions) and a table of random numbers. Nowadays (1984/5) desktop computers are further revolutionizing the ease of mathematical experimentation.

1.3. SIMULATION AS EXPERIMENTATION

We have stressed that simulation is experimental mathematics and that simulation studies should be designed carefully, a process often termed *variance reduction* in this field. Their classification as experiments also has repercussions for the reporting of simulation studies. It is essential that enough details are given for the experiments to be repeated and the results checked. Hoaglin and Andrews (1975) gave some standards on reporting which seem to have been followed only exceptionally. In view of the preceding warnings on the deficiencies of certain pseudo-random-number generators, it is important to report the generator used.

Good design is the key to reducing the cost of the study when this is necessary. The cost of generating random variables and sampling from stochastic models is usually a tiny part of the cost of the study, so the main aim should be to make best use of a small number of replications.

The analysis of simulation experiments also needs care, because the observations may not be independent. This can either occur deliberately as part of the design or because one is simulating a stochastic process through time. (The problems of analyzing observations of a simulated stochastic process apply equally to observing real processes, but this is done much less intensely.) Chapter 6 considers various ways to include dependence in the analysis or to select independent sets of observations.

1.4. SIMULATION IN INFERENCE

Simulation has recently become popular as part of statistical inference. The advantages are again the need to make fewer approximations, although interpretation may be more difficult. Monte-Carlo tests compare the data with simulated data from the supposed model. The similarity of real and simulated data provides a test of goodness-of-fit. Bootstrap methods resample from the data, using the data as a reference distribution to assess the variability or bias of an estimator. Both are discussed in Chapter 7.

1.5. EXAMPLES

Checking Distribution Theory

"Student" (1908) when deriving his t distribution carried out a small simulation experiment. He had 3000 physical measurements on humans which were known to be approximately normally distributed. These were shuffled and divided into 750 sets of (X_1, X_2, X_3, X_4). From each sample of size four the t statistic was calculated, giving 750 realizations to compare with the theoretical density. (This was done for each of two measurements.)

We can repeat this experiment with very much less effort. Figure 1.1 shows a simple BASIC program to do so. The 750 numbers can be compared with a t distribution in any way we choose. Perhaps the simplest thing to do is to compare some moments with their population values. Each run of this program on a BBC microcomputer took 130 sec. (Appendix A gives details of the computers used in this work.)

Simulation is often useful to check theoretical calculations. For example, the author was asked to check the solution to Sylvester's problem (Kendall

```
10 DIM X(4)
20 FOR I% = 1 TO 750
30 FOR J% = 1 TO 3 STEP 2
40 U = 2•RND(1) − 1
50 V = 2•RND(1) − 1
60 W = U•U + V•V
70 IF W > 1 THEN 40
80 C = SQR((−2•LN(W))/W)
90 X(J%) = C•U
100 X(J% + 1) = C•V
110 NEXT J%
120 SUM = 0
130 FOR J% = 1 TO 4
140 SUM = SUM + X(J%)
150 NEXT J%
160 XBAR = SUM/4
170 SUM = 0
180 FOR J% = 1 TO 4
190 SUM = SUM + (X(J%) − XBAR)^2
200 NEXT J%
210 S = SQR(SUM/3)
220 T = SQR(4)•XBAR/S
230 PRINT T
240 NEXT I%
```

Figure 1.1. A BASIC program to repeat Student's simulations. The function RND(1) returns a pseudo-random number. Lines 40 to 100 code algorithm 3.9 to produce normal variates.

and Moran, 1963; Solomon, 1978). Four points are placed at random in a disc and their convex hull found. What is the probability that it is a triangle? The theoretical value is $35/12\pi^2$. A simulation study was performed with 100,000 replications. In 29,432 cases the convex hull was a triangle, giving a 95% confidence interval for the probability of (0.2915, 0.2971) and confirming the theoretical value, 0.2955. The whole study took half an hour, using a VAX11/782 (including programming).

Much of statistical practice is based on asymptotic distributions, and simulation is much used to check the accuracy of asymptotic results for small samples. Ripley and Silverman (1978) considered the distribution of d, the smallest distance between any pair of n random points in the unit square. Their asymptotic result is that $n(n-1)d^2$ has an exponential distribution with mean $2/\pi$ (see also Theorem 2.6). Large values of d provide the rejection region of a test of inhibition between points, so we will count the number of values of $T = n(n-1)d^2 \geqslant 1.907$, the 95% point of the asymptotic distribution. Figure 1.2 shows the program and Table 1.1 gives the results. The count has a binomial (10,000, 0.05) distribution on the asymptotic theory, so the acceptance region of a 5% test is (457, 543) (using a normal approximation). Thus our experiment gives us no reason to doubt the asymptotic theory even for sample sizes as small as $n = 10$.

```
10 INPUT "N", N%
20 DIM X(N%), Y(N%)
30 INPUT "Reps", R%
40 CNT = 0
50 DC = 1.907/(N%•(N% – 1))
60 FOR L% = 1 TO R%
70 FOR I% = 1 TO N%
80 X(I%) = RND(1)
90 Y(I%) = RND(1)
100 NEXT I%
110 D = 2
120 FOR I% = 2 TO N%
130 X1 = X(I%): Y1 = Y(I%)
140 FOR J% = 1 TO I% – 1
150 DD = (X1 – X(J%))^2 + (Y1 – Y(J%)))^2
160 IF DD < D THEN D = DD
170 NEXT J%, I%
180 IF D > DC THEN CNT = CNT + 1
190 NEXT I%
200 PRINT "Count ="; CNT
```

Figure 1.2. BASIC program to check exponential distribution for $n(n-1)d^2$.

Table 1.1. Results from Figure 1.2

n	CNT	out of	R%	Time (min)
10	516	10,000	103
15	516	10,000	227
20	509	10,000	405

This experiment was run overnight on a personal computer and so was free. Nevertheless we should still consider whether we could have obtained more information from the experiment. [In fact we only used the fact that at least one or no pairs (x, y) had $n(n - 1)d(x, y) < 1.907$, so we could have stopped searching as soon as one was found.] Clearly we could have checked other percentage points with the same data. Could we make use of the actual values of T? One possibility is to assume that the tail of the distribution of T is exponential of unknown mean λ^{-1}, and to estimate $P(T > 1.907) = e^{-1.907\lambda}$ for an estimate $\hat{\lambda}$ of λ, say obtained from the observations with $T > 1$. Exercise 1.4 shows that this idea is worthwhile only in the extreme tail.

Comparing Estimators

Andrews et al. (1972) report a large simulation experiment that used variance reduction very effectively. Consider a location-parameter estimation problem:

$$\text{Estimate } \theta \text{ in } \{ f(x - \theta) \mid x \in \mathbb{R} \} \quad \text{from } x_1, \ldots, x_n$$

The density f is symmetric and is similar to the normal density. The idea is to find estimators that perform well across a wide class of possible densities f. Some obvious estimators of θ are the sample mean and the sample median, and a trimmed mean (the mean of all except the r largest and r smallest values). Let $T(x)$ be such an estimator. All the estimators considered were location equivariant ($x_i \to x_i + c$ implies $T \to T + c$) and many were scale equivariant ($x_i \to sx_i$ implies $T \to sT$). Our examples are both location and scale equivariant.

The key to the variance reduction was that all simulations were done for f belonging to the so-called normal/independent family. That is, f is the density of $X = Z/S$, where $Z \sim N(0, 1)$ and $S > 0$ is independent of Z. Consider first conditioning on $S_1 = s_1, \ldots, S_n = s_n$. Then $X_i \sim N(0, 1/s_i^2)$, and suitable statistics for the $\{X_i\}$ are $\hat{X}$ and $\hat{S}$ where

$$\hat{X} = \frac{\sum X_i s_i^2}{\sum s_i^2}$$

$$S^2 = \frac{\sum (X_i - \hat{X})^2 s_i^2}{n - 1}$$

Define $C_i = (X_i - \hat{X})/S$. Then for a location and scale equivariant estimator T,

$$T(\mathbf{x}) = \hat{x} + sT(\mathbf{c})$$

The point here is that $T(\mathbf{c})$ is much less variable than $T(\mathbf{x})$. We will assume T is unbiased, so $E_\theta(T) = \theta$. Consider

$$E[(T - \theta)^2 | C = \mathbf{c}, S = \mathbf{s}] = v(\mathbf{c}, \mathbf{s})$$

say, so the expectation is merely over the location and scale of the sample. Conditionally, $\hat{X}$ and S are independent, with $\hat{X} \sim N(\theta, 1/\sum s_i^2)$ and $(n - 1)S^2 \sim \chi_{n-1}^2$. Thus

$$v(\mathbf{c}, \mathbf{s}) = E\{\hat{X} - \theta + ST(\mathbf{c})\}^2$$

$$= E(\hat{X} - \theta)^2 + E(\hat{X} - \theta)ST(\mathbf{c}) + E(S^2)T(\mathbf{c})^2$$

$$= \frac{1}{\sum s_i^2} + T(\mathbf{c})^2$$

where all expectations are conditional on $C = \mathbf{c}, S = \mathbf{s}$. Finally,

$$\text{var}(T) = E\left[\frac{1}{\sum S_i^2} + T(C)^2\right]$$

and this is found by a simulation experiment as an average over many samples $(X_1, \ldots, X_n)$ of the random variables. Almost no more work is needed than in calculating $T(X)$, but the estimate of $\text{var}(T)$ obtained is much more accurate (see Table 1.2).

The essence of this transformation is to average analytically over as much of the variation as possible. The assumption on f is slightly restrictive, but includes Student's t distribution as well as the Cauchy, Laplace, and con-

Table 1.2. Estimates of $n \times \text{var}(T)$ Based on 200 Replications for Sample Size $n = 25$ for the Mean, Median, and Trimmed Mean $(r = 2)$ Estimators T

	\multicolumn{5}{c}{α}				
	1.5	2	5	10	100
Mean					
Average	1.57	1.53	1.185	1.096	1.009
s.e.1[a]	0.14	0.18	0.10	0.095	0.084
s.e.2[a]	0.048	0.062	0.019	0.010	0.0017
Variance reduction	9	8	27	90	2,400
Median					
Average	2.17	1.96	1.67	1.55	1.60
s.e.1	0.19	0.23	0.14	0.14	0.11
s.e.2	0.11	0.093	0.064	0.052	0.051
Variance reduction	3	6	5	7	5
Trimmed mean					
Average	1.72	1.61	1.22	1.14	1.051
s.e.1	0.18	0.16	0.080	0.097	0.084
s.e.2	0.065	0.065	0.022	0.015	0.0052
Variance reduction	7	6	12	40	260

[a]The s.e.1 and s.e.2 are standard errors from direct estimation and conditional estimation. The distribution of S_i^2 was $\alpha^{-1} \times \text{gamma}(\alpha)$, so $X_i \sim t_{2\alpha}$.

taminated normal distributions. It is a small price to pay for a six-fold reduction in experimental replication. It should be stressed that negligible extra work is involved. Instead of for each replication

1. Sample $Z_1, \ldots, Z_n \sim N(0, 1)$, $S_1, \ldots, S_n$, set $X_i = Z_i / S_i$
2. Form $V = T(X)^2$

and averaging V, we

1. Sample $Z_1, \ldots, Z_n \sim N(0, 1)$, $S_1, \ldots, S_n$, set $X_i = Z_i / S_i$
2. Calculate $\hat{X}, \hat{S}$
3. Form $V = 1/\sum S_i^2 + \{T(X) - \hat{X}\}^2 / S^2$

and average V. The variance reduction is most when $T(c)$ is most nearly constant, but is always worthwhile.

A Queueing Problem

Consider the following everyday queueing problem. A bank has several tellers serving customers. We could propose any one of a number of queueing disciplines:

    (i)   One common queue, with a teller on becoming free serving the customer at the head of the queue.

    (ii) Separate queues for each teller; each customer chooses the shortest queue on arrival and remains in it.

    (iii) Arriving customers choose a queue at random and remain in it.

    (iv) Customers are allocated to a queue in rotation.

    (v) Variants of (ii), (iii), and (iv) in which queue-changing is allowed if a queue becomes empty.

Any number of criteria can be used to assess the performance of the system under these disciplines. The study could look from the bank's angle and consider the time that tellers are idle, or from the customers' point of view centering on the customer's waiting time. We will not normally be interested in average waiting time, since a customer's frustration will rise more than linearly with the delay experienced.

Analytical progress is not possible for the more complex disciplines even under simplifying assumptions on the customer arrival process and the service time distribution. Some progress can be made under further approximations (Newell, 1982), but this will ignore the subtle differences between disciplines. The only practicable approach for a detailed study is simulation.

A queueing system is determined by a sequence of events through time, the moments at which any customer changes state (arrives, changes queue, is served, or departs). A simulation is, in principle, straightforward, but care is needed to ensure that all the events are simulated in the correct order. For this reason queueing systems are usually simulated in special-purpose computer languages which take care of some of the details. Queueing systems *can* be simulated in general-purpose languages, which often give more control over what is happening. Our example was simulated in BASIC on a BBC microcomputer.

The arrival process can be simulated at the outset to produce a list of arrival times. It can be any process, even one based on observations. For illustration we took a Poisson process. For each customer the departure time will be known as soon as that customer enters service, at which point the waiting time can be recorded. For convenience the service times were taken to be constant. With any of the queueing disciplines and s servers, there are $s + 1$ possible next events; a customer arrives or one of the servers completes

0	(72)
0.0–0.2	卌 卌 卌 卌
0.2–0.4	卌 卌 1111
0.4–0.6	卌 卌 卌 11
0.6–0.8	卌 卌 卌 卌
0.8–1.0	卌 卌 111
1.0–1.2	卌 卌
1.2–1.4	卌
1.4–1.6	卌
1.6–1.8	1111
1.8–2.0	卌 卌 111
2.0–2.2	卌
2.2–2.4	1
2.4–	1

Figure 1.3. Tally for waiting times in Table 1.3.

Table 1.3. Successive (Along Rows) Waiting Times for 200 Customers at a Three-Server Queueing System with Unit Service Time and Poisson Arrivals at Rate 0.4

0*	0*	0	0	0	0.148	0.080	0.603	0	0
0	0	0	0.442	0.598	0.713	0.068	0.338	0.439	0.027
0.307	0	0	0.279	0	0	0	0.011	0.639	0*
0	0	0.187	0.517	0	0.065	0.350	0.544	0	0
0	0*	0	0	0	0.332	0.055	0	0	0*
0	0	0	0	0	0	0	0	0	0
0.	0.116	0	0	0.051	0.689	0.632	0.544	1.103	1.076
0.963	0.325	0.395	0	0	0*	0	0	0.590	0.232
0.110	0.767	0.217	0	0	0	0.449	0.094	0.240	0.917
0.893	0.525	1.143	1.149	1.150	1.206	0.414	0	0	0
0.419	0.290	0.127	0.272	0.665	0.427	0.625	0.827	0.853	0.817
0.175	0	0.143	0.184	0.002	0	0*	0	0	0.409
0.456	0.444	0.854	0.954	0.771	1.363	1.091	1.124	1.769	1.462
0.483	1.124	1.007	0.917	1.200	1.124	0.857	0.609	0.617	0.671
1.317	0.701	0.184	0.821	0	0	0	0	0	0.186
0.272	0	0	0*	0	0	0.641	0.652	0	0
0	0.086	0	0.566	0.604	0.342	0.825	0.907	0.708	0.717
0.685	0.786	1.389	1.411	1.501	2.069	1.920	1.929	1.932	1.932
2.029	2.094	1.581	1.409	1.979	1.685	1.755	1.878	1.867	1.837
1.877	1.877	1.605	2.216	1.931	1.970	2.460	2.199	2.139	1.925

*Customers marked with an asterisk arrived at an empty system.

service. The times of each of these events are known, so the appropriate event can be processed and the process can be repeated.

Such a simulation will produce a series of waiting times which can be expressed as a histogram of waiting times. (See Figure 1.3.) However, there is an important point that is often overlooked. The waiting times are clearly not independent (Table 1.3), and we must not attach undue significance to the suspiciously large number of waiting times in the range 1.8–2.0. We can extract some independent events from this simulation. When a customer arrives at a completely empty system (denoted by an asterisk in Table 1.3), the future must be independent of the past (since it depends only on future arrivals and these form a Poisson process). Let us call the parts of the simulation between the starred arrivals *tours* [following Cox and Smith (1961)]. Then the tours are independent. This device has been termed *regenerative simulation* by Iglehart (although it is a method of analysis rather than simulation) and is discussed with alternative analyses in Chapter 6.

1.6. LITERATURE

There is a vast literature on simulation. As an experimental subject much that has been published has been superceded. We have made no attempt to give a comprehensive survey; bibliographies up to the late 1970s are given by Nance and Overstreet (1972), Sahai (1979), and Sowey (1972, 1978). Later papers are likely to appear either in applied or computational statistics journals or in the computing literature, with case studies in operations research and management science journals.

1.7. CONVENTIONS

The following mathematical conventions will be used in later chapters.

It is at times important to distinguish between the sequence (a_i) and the set $\{a_i\}$, the distinction being that (a_i) is ordered, whereas $\{a_i\}$ is merely the set of values taken by a sequence. For a periodic sequence (a_i) is infinite, but $\{a_i\}$ is finite.

The modulus operator $a \bmod b$ forms the remainder when a is divided by b. Mod is used in two senses: both $a \bmod b = c$ and $a \equiv c \bmod b$ to mean $a \bmod b = c \bmod b$.

The exclusive-or logical operator EOR is defined by $p\,\text{EOR}\,q$ is false if both p and q are true or both are false, and true otherwise. It can also be applied to $\{0, 1\}$, where 0 represents false and 1 true, and to binary vectors elementwise.

The function $\gcd(a, b)$ is the largest factor common to both integers a

and b. Some of the proofs of Chapter 2 will require familiarity with its use, including the existence of integers s and t such that $sa + tb = \gcd(a, b)$.

The natural logarithm is referred to as $\ln(\ )$ (except in computer programs).

Random variables have distributions described by cdf's (comulative distribution functions) and perhaps pdf's (probability density functions). We will use the notation $X \sim N(\mu, \sigma^2)$ or say X is a normal variate to indicate that the random variable X has a normal distribution. Standard distributions are assumed throughout. The gamma distribution is always assumed to have unit scale parameter and shape parameter α, that is, its pdf is

$$x^{\alpha-1}e^{-x}/\Gamma(\alpha) \quad \text{on} \quad (0, \infty)$$

The symbol $\square$ denotes the end of a proof or example.

EXERCISES

1.1. Verify that the pseudo-random sequence $1, 8, 13, \ldots$ is produced by the rule $X_i = (X_{i-1} + X_{i-2} + X_{i-3})$ mod 20, and that the sequence repeats after 248 terms. Experiment with other starting values and note that most but not all repeat after 248 terms. Consider replacing 20 first by 5, then 4, and consider all possible cases. (There are only 125 and 64 starting values, respectively.) Use Lemma A of Section 2.7 to deduce the complete solution to the original problem. [This may be easier after studying Chapter 2.]

1.2. Reflect on why models are used in your field of study, and whether simulation is as helpful for science as it is for management.

1.3. Look up Hoaglin and Andrews (1975) and compare their recommendations with simulation studies in a recent issue of a statistical journal. Do any of these studies use variance reduction?

1.4. Suppose $X_1, \ldots, X_n \sim \exp(\lambda)$, and we wish to estimate the $(1 - \alpha)$ point of their distribution. Compare the estimator $\hat{p}$ of $p = P(X_1 > c)$ obtained from the proportion of $X_i > c$ with $\tilde{p}$ obtained by fitting an exponential distribution to observations $> C$. Show that $\text{var}(\hat{p}) \approx p/n$ whereas $\text{var}(\tilde{p}) \approx (-p \ln p)^2/(ne^{-\lambda c})$. For $\lambda = \pi/2$ and $C = 1$ find the standard deviations of $\hat{p}$ and $\tilde{p}$ for $p = 10\%, 5\%,$ and 0.1%.

1.5. Show how to use the variance reduction of Andrews et al. (1972) to estimate $P(T < c)$ for c in a tail of the distribution. A greater gain is found for these probabilities than for $\text{var}(T)$.

1.6. Simulate queueing discipline (ii) (in the queueing subsection of Section 1.5) for $s = 2$ servers, unit server time, and Poisson arrivals with rate 0.6 for 250 customers. [Algorithm 3.2 shows how to generate the inter-arrival times.]

CHAPTER 2

Pseudo-Random Numbers

Almost all the simulation methods and algorithms to be discussed in later chapters derive their randomness from an infinite supply $U_0, U_1, U_2, \ldots$ of *random numbers*; that is, an independent sequence (U_i) of random variables uniformly distributed on $(0, 1)$. Many users of simulation are content to remain ignorant of how such numbers are produced, merely calling standard functions to produce them. Such attitudes are dangerous, for random numbers are the foundations of our simulation edifice, and problems at higher levels are frequently traced back to faulty foundations.

This chapter is of rather specialized appeal. Do not yield to the temptation to skip it without working exercise 2.17, for many of the random number generators in use (at the time of writing) have serious defects.

2.1. HISTORY AND PHILOSOPHY

There is no mathematical problem with random numbers: their existence is provable from Kolmogorov's axioms for probability. [See, for example, Neveu (1965, Section 5.1).] However, this result does not produce a realization of a sequence of random numbers for us to use; we have to find some observable process of which the mathematics is a reasonable model. The philosophical problem hinges on that much-abused word "reasonable." How can we decide from a finite sequence $(U_1, \ldots, U_n)$ whether (U_i) is an adequate model? We have immediately all the philosophical problems of statistical inference.

The earliest users of simulation used physical processes which were accepted as random. Most readers will have performed experiments on tossing coins or throwing dice when learning about probability. Such simulation experiments have a long history. A more sophisticated variant from the 18th century is Buffon's needle experiment to estimate π. (See also Section 7.5.) Mechanical devices are still widely used in gambling (dice, roulette wheels) and in lotteries [see West (1955) and Inoue et al. (1983)].

14

Tippett (1927) produced a table of 40,000 digits "taken at random from census reports." Later tables, such as the RAND (1955) one of a million digits, were produced from electronic noise, which is also the random input to the British "Premium Bond" draw (Thompson, 1959).

All these physical methods have been widely accepted as random, presumably on the basis of observation and explicit or implicit testing. However, many of them have been found to exhibit biases and dependencies. In the case of the RAND machine there were mechanical faults in the recording mechanism that marred the randomness suggested by the theory of electronics (Hacking, 1965, p. 129). Thus even physical devices need to be tested.

Simulation was one of the earliest uses of electronic computers. The pioneers of computing the 1940s found that physical devices did not mesh well with digital computers. Even when tables of random numbers were available on punched cards or tape, they were too slow and cumbersome. So they looked for simple ways to produce haphazard sequences, and considered various nonlinear recursive schemes. One of the earliest was von Neumann's "middle square" method. Suppose we want a sequence of four-digit decimal numbers. Starting from 8653 we square it (74874409) and extract the middle four digits, 8744. This can be repeated to obtain

$$8653, 8744, 4575, 9306, 6016, 1922, 6940, \ldots \tag{1}$$

a deterministic sequence that appears random. Hence, the terminology of *pseudo-random* (*pseudo-*: false, apparent, supposed but not real—*Concise Oxford Dictionary*) or *quasi-random* (*quasi-*: seeming, not real, half-, almost-—*Concise Oxford Dictionary*). We give a formal definition.

Definition: A sequence of *pseudo-random* numbers (U_i) is a deterministic sequence of numbers in $[0, 1]$ having the same relevant statistical properties as a sequence of random numbers.

This needs clarifying by specifying which properties are relevant and statistical. Informally, what is meant is that any statistical test applied to a finite part of (U_i) which aims to detect relevant departures from randomness would not reject the null hypothesis. In practice it seems sufficient to insist that the joint distributions of $(U_{i+1}, \ldots, U_{i+k})$ are not far from uniformity in $[0, 1]^k$ for small values of k (say, $k \leqslant 6$).

One of the most appealing ways of viewing this definition is in terms of *predictability*. In everyday speech we call things "random" if we cannot predict them. For example, one would quickly reject the output of the algorithm

$$U_i = (U_{i-1} + U_{i-2}) \bmod 1 \tag{2}$$

given in Table 2.1 when one notices that U_i never lies in (U_{i-2}, U_{i-1}) and so can be predicted to some extent. The middle square example looks unpredictable for a while, then settles down to

$$2100, 4100, 8100, 6100, 2100, \ldots$$

What we need are sequences that are hard to predict unless the mechanism generating them is known.

This introduces a connection with *cryptography*, the art or science of turning meaningful sequences into apparently random noise in such a way that a key-holder can recover the original data. The author's first acquaintance with pseudo-random numbers came in this way. A sonar device was to be constructed using pseudo-random acoustical noise. The pseudo-randomness made it unlikely that an enemy would recognize the sonar as a signal amongst oceanic noise, whereas the known structure enabled the sonar to recognize echoes of its own emissions.

Unpredictability is also the key as to why we accept physical devices as random. We know that if we had a sufficiently precise knowledge of the initial position and spin of a roulette wheel we could predict its outcome. However, the mechanism used magnifies the initial conditions to make imprecise knowledge useless for prediction. Thus we use randomness to cover our ignorance of the details of the process used, and we can do the same for nonlinear recursions.

Some Common Generators

The middle-square method and (2) were quickly rejected as sources of pseudo-random numbers, but one method of that era has survived. Lehmer (1951)

Table 2.1. Fifty Numbers from (2); read Down Columns

0.563	0.478	0.218	0.396	0.455
0.624	0.527	0.163	0.527	0.692
0.187	0.005	0.382	0.923	0.147
0.811	0.531	0.545	0.450	0.839
0.999	0.536	0.926	0.373	0.986
0.810	0.067	0.471	0.824	0.825
0.809	0.603	0.397	0.197	0.811
0.620	0.671	0.867	0.020	0.635
0.429	0.274	0.264	0.217	0.446
0.049	0.945	0.132	0.238	0.082

reported experiences with the *congruential* generator

$$U_i = aU_{i-1} \bmod 1 \qquad (3)$$

which mimics the magnification effect of a roulette wheel provided the *multiplier a* is large. In practice (3) must be computed in finite-precision arithmetic, so it is usual to generate integers X_i by

$$X_i = aX_{i-1} \bmod M \qquad (4)$$

and set $U_i = X_i/M$. Provided a and M are integers, (3) is then performed exactly. (Lehmer used $a = 23$, $M = 10^8 + 1$ on a *decimal* computer.) This family of generators and its cousins is now widespread.

Cryptographers are more concerned with pseudo-random sequences of *bits* and the use of special hardware. Thus they have tended to prefer pseudo-random number generators based on *shift-registers*. These record the last d bits $b_{i-1}, \ldots, b_{i-d}$, so

$$b_i = f(b_{i-1}, \ldots, b_{i-d})$$

for some function $f: \{0, 1\}^d \to \{0, 1\}$. The usual choice is

$$b_i = (a_1 b_{i-1} + \cdots + a_d b_{i-d}) \bmod 2 \qquad (5)$$

for binary constants $a_1, \ldots, a_d$. There are many ways to obtain pseudo-random numbers from a sequence of pseudo-random bits. The simplest is to let

$$U_i = 0.b_{iL} b_{iL+1} \cdots b_{iL+M}$$

for integers L and M with $0 < M < L$.

These generators are discussed in detail in the next three sections. It is worth noting that minor variations may give rise to very different behavior. For example, (3) implemented in floating-point arithmetic may behave quite differently from an exact implementation via (4) (cf. exercise 2.4). All these generators have the property that eventually they reach a sequence that repeats itself; the middle square example had such a sequence of length four. This length is called the *period* of the generator. Clearly four is unacceptable, and the period should be as long as possible.

There are other generators implemented on popular microcomputers, apparently without reference to the existing literature. The BASIC interpreter on the Research Machines 380Z forms

$$V_i = 38965U_{i-1} + 26664$$

and then renormalizes, $U_i = 2^\alpha V_i$, where α is chosen so that $\frac{1}{2} < U_i \leqslant 1$. This reaches a cycle of period 1995 after about 10,000 calls (Research Machines Ltd., 1982).

The BASIC interpreters on both the APPLE II and CBM PET micro-computers have used

$$V_i = 0.708076143 \times U_i$$

again renormalized so $\frac{1}{2} < W_i = 2^\alpha V_i \leqslant 1$. Now W_i is a 32-bit number, so

$$W_i = 0.B_1 B_2 B_3 B_4$$

for bytes $0 \leqslant B_i < 256$ (and $B_1 \geqslant 128$). Then on the Apple $U_i = 0.B_4 B_2 B_3 B_1$, whereas on the Pet $U_i = 0.B_4 B_3 B_2 B_1$ (Henery, 1983).

Testing

The sequences produced by all these generators do have some structure. Most of the rest of this chapter is devoted to identifying that structure and assessing its consequences. Any generator can be tested empirically by applying statistical tests for independence and uniformity to $(U_1, \ldots, U_N)$ for large N. However, this can be very time-consuming and always leaves open the possibility that there is some relevant structure which has not been detected.

For certain congruential and shift-register generators it is possible to find exactly the distribution of $(U_i, \ldots, U_{i+k-1})$ for small k. These theoretical tests have proved more searching than empirical tests. Thus, one is recommended to choose a generator for which theoretical tests are available and have been performed before it is put to serious use. (A mild amount of predictability might be an asset in an arcade-style game.)

The undesirable structure of (3) and (5) has led some authors to suggest applying further algorithms to their output in an attempt to destroy the structure. This might involve permuting the (U_i) or choosing between two or more generators for each i (See Section 2.5). Beware of the assumption that they improve matters. Very little progress has been made on their theoretical analysis, and the possibility remains that the known structure is transformed to something worse or that further structure is introduced. Complex algorithms are by no means necessarily more "random" than simple ones, as is shown by the striking example of Knuth (1981, pp. 4–5).

The conviction of this author is that it is better to use simple and well-understood algorithms, and that within families meeting those conditions it is possible to choose pseudo-random number generators good enough for any prespecified purpose.

*Random Sequences

Philosophers have discussed several ways to define randomness, but few are relevant to our purposes. The point of view taken above is close to that of Hacking (1965), who was most interested in finite sequences in attempting to understand the foundations of statistics. Another approach particularly associated with von Mises (1919, 1957) is to define probability directly in terms of limiting frequencies of infinite sequences. A *collective K* is an infinite sequence of outcomes satisfying certain conditions. The probability of an event E is defined as the limit as $n \to \infty$ of the frequency of E in the first n terms of K. Elementary texts often introduce $P(\text{coin tossed gives heads}) = \frac{1}{2}$ in this way.

Von Mises' original conditions were too strong, but they were relaxed by others and put into definitive form by Church (1940). We say (U_i) is k-distributed if the empirical distribution of $(U_i, \ldots, U_{i+k-1})$ converges to the uniform distribution on $[0, 1]^k$. Then (U_i) should be k-distributed for all k, called ∞-distributed. Furthermore, any subsequence of (U_i) should be ∞-distributed. One has to confine attention to *computable* subsequences to avoid allowing the choice of all $U_i \geqslant \frac{1}{2}$. Computable subsequences are what are known to probabilists as optional sampling rules and insist that whether or not U_{i+1} is included is determined by knowledge of $U_1, \ldots, U_i$. There exist sequences (U_i) for which all computable subsequences are ∞-distributed; these are the von Mises–Church collectives.

All our algorithms give rise to periodic sequences and so are not even 1-distributed as only a finite set of values will occur. However, only a theory of random finite sequences seems relevant to simulation. Kolmogorov (1963) had one idea, and Chaitin (1966) and Martin-Löf (1966) defined randomness in terms of the complexity of the algorithm necessary to generate the sequence.

None of these helps with our practical problem, and we will take the pragmatic approach of making (U_i) as featureless as possible; where structure is unavoidable we will aim to make its scale small.

---

*Starred subsections are optional reading.

2.2. CONGRUENTIAL GENERATORS

Congruential generators are defined by

$$X_i = (aX_{i-1} + c) \bmod M \tag{1}$$

for a *multiplier a*, *shift c*, and *modulus M*, all integers. We can and will take a, c, X_i to all be in the range $\{0, 1, \ldots, M - 1\}$. The pseudo-random sequence (U_i) is determined by (1) and

$$U_i = X_i/M \tag{2}$$

once the *seed* X_0 is given. We saw in Section 2.1 that generators of this form with $c = 0$ were first described by Lehmer (1951); such generators are called *multiplicative*. The early literature contains some confusion about the general (*mixed*) case; the first example published seems to be $a = 2^7, c = 1, M = 2^{35}$ by Rotenburg (1960).

The future of (X_i) is determined by its current value. Since the $M + 1$ values $(X_0, \ldots, X_M)$ cannot be distinct, at least one value must occur twice, as X_i and X_{i+k}, say. Then $X_i, \ldots, X_{i+k-1}$ is repeated as $X_{i+k}, \ldots, X_{i+2k-1}$, and so the sequence (X_i) is periodic with a period $k \leqslant M$. The *full period M* can always be achieved with $a = c = 1$. Table 2.2 illustrates the range of behavior that can occur. Clearly, the period depends on the choice of a, c and perhaps also on the seed. For multiplicative generators the *maximal* period is $M - 1$, for if 0 ever occurs it is repeated indefinitely.

It is usual to choose M to make the modulus operation efficient, and then to choose a and c to make the period as long as possible. It is known how to find the period of an arbitrary congruential generator (Fuller, 1976; Dudewicz and Ralley, 1981) but this seems unnecessary as the following theorems suffice. Proofs are given in Section 2.7.

Theorem 2.1. A congruential generator has full period M if and only if

 (i) $\gcd(c, M) = 1$.
 (ii) $a \equiv 1 \bmod p$ for each prime factor p of M.
 (iii) $a \equiv 1 \bmod 4$ if 4 divides M.

Note that if M is a prime, full period is attained only if $a = 1$.

Theorem 2.2. A multiplicative generator with modulus $M = 2^\beta \geqslant 16$ has maximal period $M/4$, attained if and only if $a \bmod 8$ is 3 or 5. In the case $a \equiv 5 \bmod 8$, let $b = X_0 \bmod 4$. Then $(U_i - b/M)$ is the sequence output

Table 2.2. Examples of Congruential Generators

(a)	$M = 16, a = 1, c = 1$
	0, 1, 2, 3, 4, 5, 6, 7, 8, 9, 10, 11, 12, 13, 14, 15, 0, ...
(b)	$M = 16, a = 5, c = 1$
	0, 1, 6, 15, 12, 13, 2, 11, 8, 9, 14, 7, 4, 5, 10, 3, 0, ...
(c)	$M = 16, a = 5, c = 4$
	0, 4, 8, 12, 0, ... or 1, 9, 1, ... or 2, 14, 10, 6, 2, ... or 3, 3, ...
	or 5, 13, 5, ... or 7, 7, ... or 11, 11, ... or 15, 15, ...
(d)	$M = 16, a = 5, c = 0$
	1, 5, 9, 13, 1, ... or 2, 10, 2, ... or 3, 15, 11, 7, 3, ... or
	4, 4, ... or 6, 14, 6, ... or 8, 8, ... or 12, 12, ...
(e)	$M = 16, a = 3, c = 0$
	1, 3, 9, 11, 1, ... or 2, 6, 2, ... or 4, 12, 4, ... or
	5, 15, 13, 7, 5, ... or 8, 8, ... or 10, 14, 10, ...
(f)	$M = 16, a = 4, c = 0$
	1, 4, 0, 0, ... or 2, 8, 0, 0, ..., etc.
(g)	$M = 13, a = 2, c = 0$
	1, 2, 4, 8, 3, 6, 12, 11, 9, 5, 10, 7, 1, ...
(h)	$M = 13, a = 4, c = 0$
	1, 4, 3, 12, 9, 10, 1, ... or 2, 8, 6, 11, 5, 7, 2, ...
(i)	$M = 13, a = 5, c = 0$
	1, 5, 12, 8, 1, ... or 2, 10, 11, 3, 2, ... or 4, 7, 9, 6, 4, ...
(j)	$M = 13, a = 12, c = 0$
	1, 12, 1, ... or 2, 11, 2, ... or 3, 10, 3, ..., etc.

from the full-period generator

$$X_i = \{aX_{i-1} + b(a - 1)/4\} \bmod M/4$$

The final assertion is due to Thompson (1958) and contradicts an earlier folklore that mixed generators were somehow less random than multiplicative ones. The sole advantage of a multiplicative generator seems to be to avoid $U_i = 0$.

Theorem 2.3. A multiplicative generator has period $M - 1$ only if M is prime. Then the period divides $M - 1$, and is $M - 1$ if and only if a is a *primitive root*, that is, $a \neq 0$ and $a^{(M-1)/p} \not\equiv 1 \bmod M$ for each prime factor p of $M - 1$.

Thus prime moduli are much more useful for multiplicative generators.

It may not be easy to find a primitive root, but once one is found all the others follow from:

Theorem 2.4. If a is a primitive root for a prime M, so is $a^k \bmod M$ provided $\gcd(k, M - 1) = 1$.

One example of the use of this theorem is the Mersenne prime $2^{31} - 1$. We know $2^{31} - 2 = 2.3^2.7.11.31.151.331$ from which one can check that 7 is a primitive root and hence so is $7^5 = 16807$, recommended by Lewis et al. (1969) and Gustavson and Liniger (1970).

*Reduction Modulo M

The modulus M is usually chosen to make it easy to implement $Y \bmod M$ without division. If the computer works to base r and $M = r^\beta$, all we have to do is to retain the bottom β digits of Y. For example, $12345678 \bmod 10^5 = 45678$. Thus one finds powers of 2 used on binary computers and powers of 10 used on calculators.

It is only a little more difficult to implement $M = r^\beta - s$ for small s. Let $Z = Y \bmod M$ and $z = Y \bmod r^\beta$ (which is easy). Then

$$Y = z + tr^\beta = tM + (z + st)$$

so $Z = (z + ts) \bmod M$, which can be performed by subtracting M from $z + ts$ until the result is less than M. Exercise 2.7 shows s subtractions will suffice.

More care is needed to implement a multiplicative generator with $M = r^\beta + 1$. This takes the r^β values $\{1, \ldots, r\}$. If $Y = aX_{i-1}, Y = z + tr^\beta$ and $(aX_{i-1}) \bmod M = (z - t) \bmod M = z - t$ if $z \geqslant t$, otherwise $z - t + M$. The one remaining problem is that we will have to arrange to store the value r^β as 0 in a β-digit word.

These tricks have proved quite popular. The Mersenne prime $2^{31} - 1$ has been used with several multipliers; Lehmer originally used the prime $10^8 + 1$ and the prime $2^{16} + 1$ has been popular more recently. For $r^\beta + 1$ it is usual to let $U_i = X_i/r^\beta$, to avoid a time-consuming division.

Choosing a Generator

Thus far we have restricted our choice by choosing M so that mod M is easy, and a and c to achieve full or maximal period. There is still a lot of freedom left! We will now confine attention to full period generators and multiplicative generators with a prime modulus and maximal period. These take values evenly spaced in $[0, 1)$, each occurring once per cycle. Provided M is sufficiently large, the (U_i) will be nearly uniformly distributed. They may, however, be completely predictable, for example if $a = c = 1$. A less obvious example is the once very popular generator RANDU with $M = 2^{31}$, $a = 2^{16} + 3, c = 0$. Then

$$X_{i+2} = (2^{16} + 3)X_{i+1} + c_1 2^{31} = (2^{16} + 3)^2 X_i + c_1 2^{31}(2^{16} + 3) + c_2 2^{31}$$

$$= (6.2^{16} + 9)X_i + \{(2^{16} + 3)c_1 + c_2 + 2X_i\}2^{31}$$

$$= 6(2^{16} + 3)X_i - 9X_i + c_3 2^{31}$$

$$= 6X_{i+1} - 9X_i + c_4 2^{31}$$

where each c_i is an integer. Thus

$$U_{i+2} - 6U_{i+1} + 9U_i \text{ is an integer}$$

and (U_i, U_{i+1}, U_{i+2}) lies on one of 15 planes in the unit cube. This means that if (U_{i-1}, U_{i-2}) is known even to limited accuracy, then U_i is quite predictable. This is a fairly extreme example, but it has been very widely used on IBM 360/370 and PDP-11 machines!

The remedy is to choose a and c to avoid this happening. Marsaglia (1968) pointed out and Fig. 2.1 demonstrates that the k-tuples $(U_i, \ldots, U_{i+k-1})$

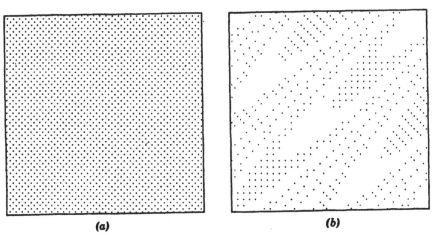

(a)   (b)

Figure 2.1. Plots of pairs (U_i, U_{i+1}) for various congruential generators modulo 2048. (a) $a = 65$, $c = 1$, all 2048 points. (b) First 512 points of (a) with $X_0 = 0$. (c) $a = 1365, c = 1$. (d) $a = 1229$, $c = 1$. (e) $a = 157, c = 1$. (f) $a = 45, c = 0$. (g) $a = 43, c = 0$.

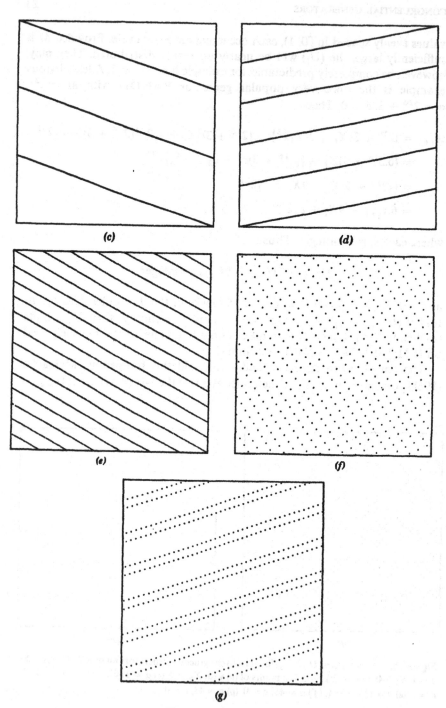

Figure 2.1 (Continued)

will always lie on a finite number of hyperplanes in $[0, 1)^k$. More precisely, we have (Beyer et al., 1971, Smith, 1971, Ripley 1983a).

Theorem 2.5. Let Λ_k be the lattice $\{t_1 e_1 + \cdots + t_k e_k \mid t_i$ integer$\}$ with basis

$$e_1 = \frac{1}{N}(1, a, a^2, \ldots)^T \tag{3}$$

$$e_j = j\text{th unit vector for } 2 \leqslant j \leqslant k$$

Then

(i) For full-period generators

$$\{(U_i, \ldots, U_{i+k-1})\} = [0, 1)^k \cap \{(U_0, \ldots, U_{k-1}) + \Lambda_k\}$$

with $N = M$, and $\{(U_{ki}, \ldots, U_{ki+k-1})\}$ is of the same form with $N = M/\gcd(k, M)$.

(ii) For a maximal-period multiplicative generator with prime modulus M,

$$\{(U_i, \ldots, U_{i+k-1})\} = (0, 1)^k \cap \Lambda_k$$

with $N = M$, and

$$\{(U_{ki}, \ldots, U_{ki+k-1})\} = (0, 1)^k \cap \Lambda_k$$

if $\gcd(k, M - 1) = 1$, otherwise it is a nonlattice subset of $(0, 1)^k \cap \Lambda_k$.

PROOF.   See Section 2.7.   □

Figure 2.1 shows what happens in some other cases. Note that we can deduce the behavior of multiplicative generators with $M = 2^\beta$, $a \equiv 5 \bmod 8$ from the full-period case by Theorem 2.2; one merely replaces M by $M/4$ in defining Λ_k.

One important conclusion of Theorem 2.5 is that the choice of c merely shifts the lattice. It has been traditional that c be chosen to minimize the correlation between U_i and U_{i+1}. However, for a fine lattice like Fig. 2.1a the correlation will be small for any c, whereas for Fig. 2.1c to minimize the correlation will merely mask the lack of independence of (U_i, U_{i+1}). There seems no compelling advantage of any other value over $c = 1$.

This reduces our choice to a few values of M and to the choice of a. In the Section 2.4 we show that a can be chosen to make the lattice of values described in Theorem 2.5 rather evenly spread in $[0, 1)^k$ and thus about $N^{-1/k}$ apart. $\{[0, 1)^k \cap \Lambda_k$ contains N points.$\}$ We would like this distribution to be as even as possible, which means choosing M as large as possible. An idea of how large M might need to be can be obtained from:

Theorem 2.6. Suppose n points are uniformly and independently distributed in $(0, 1)^k$. Let D be the smallest distance between a pair of points. Then D^k is approximately exponentially distributed with mean $2\Gamma(\tfrac{1}{2}k + 1)/\pi^{k/2}n^2 = \alpha_k/n^2$, say.

PROOF. Ripley (1983a). The approximation is asymptotic as $n \to \infty$ but remarkably accurate for n as small as 25. □

Suppose our simulations need n k-tuples. Then provided $N^{-1/k} \leqslant (\alpha_k/99.50n^2)^{1/k}$, the 1% point of the distribution of D, the nonuniformity of the pseudo-random numbers will be negligible. This reduces to $N \geqslant 200n^2$, say 2^{27} for $n = 1000$. Thus:

Recommendation. A congruential generator should have period as large as possible, at least 2^{30}, a multiplier a chosen to give period M or $M - 1$, *and* a good lattice structure as described in Section 2.4.

2.3. SHIFT-REGISTER GENERATORS

Shift registers were introduced in Section 2.1. In their most general form they have $M \geqslant 2$ states, but we will confine attention to the binary case which has been the only one used for pseudo-random numbers. We have

$$b_i = (a_1 b_{i-1} + \cdots + a_d b_{i-d})\bmod 2 \qquad (1)$$

This is easy to implement in a hardware circuit by use of a shift register, hence the name. Note that addition modulo 2 and exclusive or have the same truth table, and so we may replace (1) by

$$b_i = b_{i-j_1} \text{ EOR } b_{i-j_2} \cdots b_{i-j_k}$$

where $a_{j_1} = \cdots = a_{j_k} = 1$ and all other $a_j = 0$.

Each b_i is determined by $(b_{i-1}, \ldots, b_{i-d})$, which has at most 2^d possible values. Furthermore, if this is the zero vector, then $b_i = 0$, and $b_j = 0$ for all $j \geqslant i$. Thus, the maximal period is $2^d - 1$. The details of how to find the period of (1) (or even if the maximal period is attained) depend on methods of factorizing polynomials over finite fields. Golomb (1967) summarizes the algebra needed. Recursion (1) is associated with the polynomial

$$f(x) = x^d + a_1 x^{d-1} + \cdots + a_d$$

It has been usual to consider trinomials $1 + x^q + x^p$ with $1 \leq q < p$, so

$$b_i = b_{i-p} \text{ EOR } b_{i-(p-q)} \qquad (2)$$

Reversing the sequence shows $1 + x^{p-q} + x^p$ and

$$b_i = b_{i-p} \text{ EOR } b_{i-q}$$

must have the same period. Table 2.3 lists some pairs (p, q) that give maximal period $2^p - 1$. [From Golomb (1967). Further values are given by Lewis and Payne (1973, Fig. 9), Zierler and Brillhart (1968, 1969), and Zierler (1969).] Some specific suggestions are $p = 98$, $q = 27$ (Lewis and Payne, 1973); $p = 521$, $q = 32$ (Bright and Enison, 1979); and $p = 607$, $q = 273$ (Tootill et al., 1973).

Tausworthe (1965) suggested using

$$U_i = \sum_1^L 2^{-s} b_{it+s} = 0.b_{it+1} \cdots b_{it+L}$$

that is, L-bit binary fractions taken t apart. Consequently, such random-number generators are called Tausworthe generators. The parameter t is called the *decimation*. A decimation is said to be *proper* if $\gcd(t, 2^p - 1) = 1$. For a proper decimation (U_i) has period $2^p - 1$ (since this is the period of each of its bits by Lemma C of Section 2.7).

The BBC microcomputer has a Tausworthe generator with $p = 33$, $q = 13$, $t = L = 32$. This is a proper decimation, and so has period $2^{33} - 1$. (The order of the *bytes* in U_i is reversed, but this has no consequence.) The following algorithm is a neat way to implement a Tausworthe generator with

Table 2.3. All Values of (p, q) for which $1 + x^q + x^p$ gives a Maximum-Period Shift Register, with $p \leq 36$

p	q	p	q	p	q
2	1	11	2, 9	25	3, 7, 18, 22
3	1, 2	15	1, 4, 7, 8, 11, 14	28	3, 9, 13, 15, 19, 25
4	1, 3	17	3, 5, 6, 11, 12, 14	29	2, 27
5	2, 3	18	7, 11	31	3, 6, 7, 13, 18,
6	1, 5	20	3, 17		24, 25, 28
7	1, 3, 4, 6	21	2, 19	33	13, 20
9	4, 5	22	1, 21	35	2, 33
10	3, 7	23	5, 9, 14, 18	36	11, 25

$q \leqslant p/2$, $p = t = L$ = word length. [It is used with $p = 36$, $q = 11$ on the Honeywell Multics system (Sibson, 1984).]

Algorithm 2.1 (Whittlesey, 1968; Payne 1970). Assume U_i is stored in a word X with b_{it+1} on the left:

1. Copy X to T.
2. Left shift X by q bits, filling with zeroes.
3. Let $X = X$ EOR T, copy X to T (bitwise exclusive or).
4. Right shift T by $p - q$ bits, filling with zeroes.
5. $X = X$ EOR T now contains U_{i+1}.

One can use $p = t = L$ < word length by padding with zeroes on the right. Exercise 2.9 shows that this algorithm works.

Lewis and Payne suggested making up an L-bit integer from nonconsecutive terms in (b_i), for example,

$$Y_i = b_i b_{i-l_2} \cdots b_{i-l_L} \tag{3}$$

for delays $l_2, \ldots, l_L$. Each bit of Y_i still obeys (2), so we can form

$$Y_i = Y_{i-p} \text{ EOR } Y_{i-(p-q)} \tag{4}$$

which can be implemented by a simple circular buffer. Such generators are called *generalized feedback shift registers* (GFSRs). They were introduced to be faster than Tausworthe generators, but Algorithm 2.1 may be faster for $p = t = L$. The p starting values for recursion (4) need not satisfy (3). However, the period of (Y_i) will depend on the starting values. Obviously we will obtain random numbers by $U_i = 2^{-L} Y_i$, so $0 < U_i < 1$.

Example. $p = 5, q = 2$ gives the bit sequence

$$1111100011011101010000100101100\ldots$$

of maximum period 31. The Tausworthe sequence with $t = L = 5$ is

31, 3, 14, 20, 4, 22, 15, 17, 23, 10, 2, 11, 7, 24, 27, 21, 1, 5, 19, 28, 13, 26, 16, 18, 25, 30, 6, 29, 8, 9, 12, . . .

If we take the GFSR $b_i b_{i-6} b_{i-12} b_{i-18} b_{i-24}$ we obtain

1, 13, 8, 29, 30, 9, 16, 22, 20, 14, 31, 4, 24, 11, 10, 7, 15, 18, 12, 5, 21, 3, 23, 25, 6, 2, 26, 17, 27, 28, 19, . . .   □

The theoretical analysis of Tausworthe and GFSR sequences concentrates on the k-tuples of integers $(Y_i, \ldots, Y_{i+k-1})$. We would like all k-tuples to be equally frequent in a period. There is a minor problem with the missing p-fold zero in (b_i). We say (Y_i) is k-*distributed* if all k-tuples are equally frequent except zero, which occurs one less time in each period.

Theorem 2.7. A Tausworthe generator with proper decimation is k-distributed for $1 \leqslant k \leqslant \text{int}[p/t]$.

PROOF. $(Y_i, \ldots, Y_{i+k-1})$ is made up from kL bits of $(b_{it+1}, \ldots, b_{it+kt})$. If $kt \leqslant p$, this is a subset of $(b_{it+1}, \ldots, b_{it+p})$ that takes all possible values except all zeroes once in a period. Thus every nonzero k-tuple of Y_i's is equally frequent in a period. $\square$

Suppose $kL > p$. Then only $2^p - 1$ of the possible $2^{kL} - 1$ values of $(Y_i, \ldots, Y_{i+k-1})$ can occur. Consequently, k-distribution is impossible for $k > \text{int}[p/L]$. Figure 2.2 shows that there may be advantages in taking $t > L$ to improve the k-dimensional structure, and that when k-distribution fails, it can fail dramatically.

The analogue of Theorem 2.7 is not automatic for GFSRs; it depends on the starting values. Let A be the $p \times L$ matrix whose rows are the bits of $Y_1, \ldots, Y_p$, called the *seed matrix*.

Theorem 2.8. A GFSR sequence is 1-distributed if and only if its seed matrix is nonsingular.

PROOF. Let A_i be the corresponding matrix for $(Y_i, \ldots, Y_{i+p-1})$. Define a $p \times p$ matrix C by

$$C_{ij} = \delta_{i,j-1} \text{ for } 1 \leqslant i < p, 1 \leqslant j \leqslant p$$

$$C_{pj} = \delta_{1j} + \delta_{qj} \text{ for } 1 \leqslant j \leqslant p$$

so $A_i = CA_{i-1} = C^{i-1}A$ with addition modulo 2.

Now $(b_{i+j}, \ldots, b_{i+j+p-1})^T = C^j(b_i, \ldots, b_{i+p-1})^T$ so $C^0, \ldots, C^s, s = 2^p - 2$, are distinct matrices; hence Y_i is 1-distributed if and only if A is nonsingular. $\square$

Theorem 2.9. A GFSR sequence is k-distributed if and only if both $k \leqslant \text{int}[p/L]$ and the matrix with row i, the bits of $(Y_i, \ldots, Y_{i+k-1})$, $i = 1, \ldots, p$ is nonsingular.

PROOF. Apply theorem 2.8 to the kL-bit integers made up by concatenating $Y_i, \ldots, Y_{i+k-1}$. $\square$

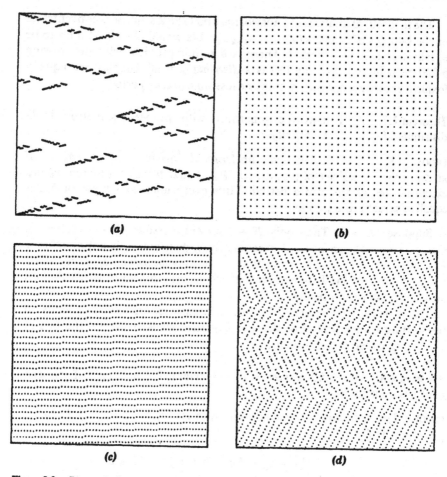

Figure 2.2. Plots of all pairs (U_i, U_{i+1}) from shift-register generators with $p = 11$, $q = 2$. (a) $t = L = 11$. (b) $t = L = 5$, so 2-distributed. (c) $t = 5$, $L = 8$. (d) $t = 17$, $L = 8$.

Example. Consider $p = 7$, $q = 1$ with period 127. A GFSR sequence with $L = 3$ gives

```
0123456131717137226624
50404615444274100653
51063664165502573052
72435755676220311402
32054211251630374753
34332607701467071521
7764736...
```

which is 2-distributed. The seed matrix is

$$\begin{bmatrix} 0 & 0 & 0 \\ 0 & 0 & 1 \\ 0 & 1 & 0 \\ 0 & 1 & 1 \\ 1 & 0 & 0 \\ 1 & 0 & 1 \\ 1 & 1 & 0 \end{bmatrix}$$

which is clearly nonsingular. For 2-distribution we consider

$$\begin{bmatrix} 0 & 0 & 0 & 0 & 0 & 1 \\ 0 & 0 & 1 & 0 & 1 & 0 \\ 0 & 1 & 0 & 0 & 1 & 1 \\ 0 & 1 & 1 & 1 & 0 & 0 \\ 1 & 0 & 0 & 1 & 0 & 1 \\ 1 & 0 & 1 & 1 & 1 & 0 \\ 1 & 1 & 0 & 0 & 0 & 1 \end{bmatrix}$$

which is nonobviously of rank six. (See below.) If we were to start with 6, 5, 4, 3, 2, 1, 0 we would obtain a nonsingular seed matrix and so 1-distribution. However, for 2-distribution we have

$$\begin{bmatrix} 1 & 1 & 0 & 1 & 0 & 1 \\ 1 & 0 & 1 & 1 & 0 & 0 \\ 1 & 0 & 0 & 0 & 1 & 1 \\ 0 & 1 & 1 & 0 & 1 & 0 \\ 0 & 1 & 0 & 0 & 0 & 1 \\ 0 & 0 & 1 & 0 & 0 & 0 \\ 0 & 0 & 0 & 0 & 1 & 1 \end{bmatrix} \tag{5}$$

which is singular since the sum (mod 2) of columns 2, 5, and 6 is zero. Thus this sequence is not 2-distributed. (See Exercise 2.11.) □

Theorems 2.7 and 2.9 show the need for very long periods. For example, Bright and Enison (1970) and Fushimi and Tezuka (1983) both consider $p = 521$, $q = 32$, with $L = 64$ and $L = 32$, respectively. By Theorem 2.9 these are candidates to be 8-distributed and 16-distributed, and Fushimi and Tezuka check that this is so. However, even with a period of $2^{521} - 1$ the

k-tuples have a spacing of 2^{-32} in dimensions 1–16, whereas congruential generators of similar periods will do much better for $k \ll \text{int}[p/L]$.

In general a very long period will be easier to achieve with a GFSR generator than a Tausworthe or congruential generator. All one has to do is to increase the size of the buffer retaining $Y_{i-1}, \ldots, Y_{i-p}$ when increasing p. One then uses the following algorithm.

Algorithm 2.2. Locations $Y[1] \cdots Y[p]$ are set aside as a buffer, initialized with $Y_{-1}, \ldots, Y_{-p}$, and pointers I and J set to $p - q$ and p

1. $Y = Y[I]$ EOR $Y[J]$, $Y[J] = Y$.
2. $I = I - 1$; if $I = 0$ then $I = p$.
3. $J = J - 1$; if $J = 0$ then $J = p$.
4. Return Y.

Normally each Y will be held within a computer word; if this is not possible, operation 1 is applied to each part of Y independently.

There remains the problem of choosing the starting values to achieve maximal k-distribution. Trial-and-error checking the conditions of Theorem 2.9 seems to be the only general way known. To achieve 1-distribution is easy; merely including $1, 2, \ldots, 2^{L-1}$ in $(Y_1, \ldots, Y_p)$ ensures that the seed matrix is nonsingular. Nonsingularity of a $p \times kL$ binary matrix is easily checked by reducing it to upper triangular form by exclusive-oring rows. For example, consider (5). The following process consists of exclusive-oring each row in turn with lower rows to remove 1's from the next column, or permuting rows. One rapidly finds the matrix to be of rank 5 and so singular.

$$
\begin{bmatrix}
1 & 1 & 0 & 1 & 0 & 1 \\
1 & 0 & 1 & 1 & 0 & 0 \\
1 & 0 & 0 & 0 & 1 & 1 \\
0 & 1 & 1 & 0 & 1 & 0 \\
0 & 1 & 0 & 0 & 0 & 1 \\
0 & 0 & 1 & 0 & 0 & 0 \\
0 & 0 & 0 & 0 & 1 & 1
\end{bmatrix} \rightarrow
\begin{bmatrix}
1 & 1 & 0 & 1 & 0 & 1 \\
0 & 1 & 1 & 0 & 0 & 1 \\
0 & 1 & 0 & 1 & 1 & 0 \\
0 & 1 & 1 & 0 & 1 & 0 \\
0 & 1 & 0 & 0 & 0 & 1 \\
0 & 0 & 1 & 0 & 0 & 0 \\
0 & 0 & 0 & 0 & 1 & 1
\end{bmatrix} \rightarrow
\begin{bmatrix}
1 & 1 & 0 & 1 & 0 & 1 \\
0 & 1 & 1 & 0 & 0 & 1 \\
0 & 0 & 1 & 1 & 1 & 1 \\
0 & 0 & 0 & 0 & 1 & 1 \\
0 & 0 & 1 & 0 & 0 & 0 \\
0 & 0 & 1 & 0 & 0 & 0 \\
0 & 0 & 0 & 0 & 1 & 1
\end{bmatrix} \rightarrow
$$

$$
\begin{bmatrix}
1 & 1 & 0 & 1 & 0 & 1 \\
0 & 1 & 1 & 0 & 0 & 1 \\
0 & 0 & 1 & 1 & 1 & 1 \\
0 & 0 & 0 & 0 & 1 & 1 \\
0 & 0 & 0 & 1 & 1 & 1 \\
0 & 0 & 0 & 1 & 1 & 1 \\
0 & 0 & 0 & 0 & 1 & 1
\end{bmatrix} \rightarrow
\begin{bmatrix}
1 & 1 & 0 & 1 & 0 & 1 \\
0 & 1 & 1 & 0 & 0 & 1 \\
0 & 0 & 1 & 1 & 1 & 1 \\
0 & 0 & 0 & 1 & 1 & 1 \\
0 & 0 & 0 & 0 & 1 & 1 \\
0 & 0 & 0 & 1 & 1 & 1 \\
0 & 0 & 0 & 0 & 1 & 1
\end{bmatrix} \rightarrow
\begin{bmatrix}
1 & 1 & 0 & 1 & 0 & 1 \\
0 & 1 & 1 & 0 & 0 & 1 \\
0 & 0 & 1 & 1 & 1 & 1 \\
0 & 0 & 0 & 1 & 1 & 1 \\
0 & 0 & 0 & 0 & 1 & 1 \\
0 & 0 & 0 & 0 & 0 & 0 \\
0 & 0 & 0 & 0 & 1 & 1
\end{bmatrix} \rightarrow
$$

$$\begin{bmatrix} 1 & 1 & 0 & 1 & 0 & 1 \\ 0 & 1 & 1 & 0 & 0 & 1 \\ 0 & 0 & 1 & 1 & 1 & 1 \\ 0 & 0 & 0 & 1 & 1 & 1 \\ 0 & 0 & 0 & 0 & 1 & 1 \\ 0 & 0 & 0 & 0 & 1 & 1 \\ 0 & 0 & 0 & 0 & 0 & 0 \end{bmatrix} \rightarrow \begin{bmatrix} 1 & 1 & 0 & 1 & 0 & 1 \\ 0 & 1 & 1 & 0 & 0 & 1 \\ 0 & 0 & 1 & 1 & 1 & 1 \\ 0 & 0 & 0 & 1 & 1 & 1 \\ 0 & 0 & 0 & 0 & 1 & 1 \\ 0 & 0 & 0 & 0 & 0 & 0 \\ 0 & 0 & 0 & 0 & 0 & 0 \end{bmatrix}$$

Another method is to select the initial values of the form (3). Then the bits of $(Y_i, \ldots, Y_{i+k-1})$ are $\{b_{i-j+k-1} | j = t, l_2 + t, \ldots, l_L + t; t = 0, \ldots, k - 1\}$. Provided this set of values is distinct, the proof of Theorem 2.7 shows that these k-tuples are 1-distributed and hence that (Y_i) is k-distributed. One must then choose the delays at least k apart and with $l_L \leqslant p - k$, which is always possible for $kL \leqslant p$, for example by

$$Y_i = b_i b_{i-k} \ldots b_{i-(L-1)k}$$

Specific implementations of generators of this type are considered by Arvillias and Maritsas (1978) and Fushimi and Tezuka (1983).

Fellen (1969) and Toothill et al. (1971, 1973) study less relevant properties of Tausworthe generators.

2.4. LATTICE STRUCTURE

We saw in Section 2.2 that the k-tuples $(U_i, \ldots, U_{i+k-1})$ from certain congruential generators lie on lattices in the unit hypercube. Both congruential and shift-register generators suffer from the same problem: for a period of length N there are only N k-tuples. For shift-register generators with k-distribution the word length L is restricted, so that these N points lie on the cubic lattice of side 2^{-L}, with $(2^{-L})^k \leqslant N + 1$. Figure 2.2 shows what happens if we increase the word length. Congruential generators can achieve a very similar k-dimensional behaviour, *provided* that the multiplier is chosen suitably. The rest of this section is devoted to a detailed study of the k-dimensional output of full-period congruential generators, and those of maximal period with a prime modulus.

Lattices
A lattice Λ in $\mathbf{R}^k$ is defined by k linearly independent vectors $e_1, \ldots, e_k$. Then

$$\Lambda = \{t_1 e_1 + \cdots + t_k e_k \mid t_i \text{ integer}\}$$

is the set of sums of integer multiples of the e_i. The set $\{e_i\}$ is called a *basis* for Λ.

Various measures of the "granularity" of a lattice have been developed. For a cubic lattice we use the smallest spacing between a pair of points, l_1, which is also the length of the smallest nonzero vector in Λ. Most people envisage a lattice as being made up by repeating a basic parallelogram cell. (Look at Fig. 2.1 again to convince yourself.) One way to define such a cell is to take e_1 as a shortest nonzero vector in Λ, e_2 as the shortest vector linearly independent of e_2, e_3 as the shortest linearly independent of e_1 and e_2, and so on. For $k \leqslant 4$ this generates a basis for Λ (except for one exceptional lattice for $k = 4$, for which only some of the choices for shortest work). However, one's intuition about lattices fails for $k \geqslant 5$. Let l_i be the length of e_i chosen in this way. Then l_k is one measure of "granularity," and $r = l_k/l_1$ measures the "uniformity" of the lattice. (We can usually achieve $r \leqslant 2$.)

Yet another method of measuring uniformity was given in Section 2.2 where we saw that the triples from RANDU lie on only 15 planes. It has proved more useful to measure the maximal spacing between parallel planes that cover the lattice. Clearly Fig. 2.1a will have a smaller spacing than Fig. 2.1c. Call this spacing s_k and its reciprocal v_k.

Computing Lattice Constants

The theory behind the following methods is described later in this section. The case $k = 2$ is easiest.

Theorem 2.10. Start with any basis (e, f) for Λ_2. Relabel if necessary so that $\|e\| \leqslant \|f\|$. Compute $s = \text{nint}(e^T f/\|e\|^2)$. If $s \neq 0$, replace f by $f - se$ and repeat. If $s = 0$, then $l_1 = \|e\|$, $l_2 = \|f\|$, and $v_2 = Nl_1$.

Remarks. (i) nint(x) is the nearest integer to x, halves being rounded toward zero, so nint(-3.5) = -3, for example. (ii) The coordinates of all vectors in Λ_2 are multiples of $1/N$ (since this is true of the basis 2.3). Thus it may be convenient to perform the calculations on Ne and Nf.

PROOF. (i) $\|f - se\|^2 = \|f\|^2 + s^2\|e\|^2 - 2se^T f < \|f\|^2$ if and only if $s \neq 0$ and $2e^T f > s\|e\|^2$, if and only if $s \neq 0$. Thus the algorithm strictly reduces the length of f and by remark (ii) must terminate.

(ii) Clearly $f - se \in \Lambda_2$, and $(e, f - se)$ is another basis.

(iii) Now suppose we have a basis with $s = 0$. By replacing f by $-f$ if necessary, we may assume $0 \leqslant e^T f \leqslant \frac{1}{2}\|e\|^2$. Suppose $g \in \Lambda_2$ and $\|g\| < \|f\|$. Then there are integers u and v with $g = ue + vf$, and by changing the sign we may assume $u > 0$. Then if $\alpha = v/u$,

$$\|g\|^2 = u^2(\|e\|^2 + \alpha^2\|f\|^2 + 2\alpha e^T f)$$

Thus if $v > 0$ we can find a shorter g by replacing v by $-v$. Then

$$\|f\|^2 > \|g\|^2 \geqslant \|e\|^2 + \alpha^2\|f\|^2 + 2\alpha e^T f$$
$$\geqslant (1 + \alpha)\|e\|^2 + \alpha^2\|f\|^2$$

This implies $-1 < \alpha \leqslant 0$, or $0 \leqslant -v < u$. Suppose $v \neq 0$. Let $\beta = -u/v > 1$. Then

$$\|f\|^2 > \|g\|^2 \geqslant \|f - \beta e\|^2 \geqslant \beta(\beta - 1)\|e\|^2 + \|f\|^2$$

a contradiction. Thus the only vectors shorter than f in Λ_2 are integer multiples of e, so $l_1 = \|e\|$, $l_2 = \|f\|$.

(iv) Choose H as the line through the origin of a family of parallel lines with spacing s_2. Choose f as a shortest vector in $H \cap \Lambda_2$. Then the basic parallelogram has base f and height s_2, so area $= s_2\|f\| = 1/N$. Thus, $s_2 = 1/N\|f\| \geqslant 1/Nl_1$, and equality is attained if H contains a vector attaining l_1. Thus $v_2 = 1/s_2 = Nl_1$. $\qquad\square$

This method of changing basis was proposed by Beyer et al. (1971) and Marsaglia (1972) but has a long history in number theory. It normally works extremely rapidly.

Example. For Fig. 2.1f we have $N = 512$, $e_1 = (1, 45)/512$, and $e_2 = (0, 1)$. For ease of working we multiply both vectors by 512.

(i) $\quad e = (1, 45)$, $f = (0, 512)$ gives $s = 11$, $f \rightarrow (-11, 17)$.

(ii) $\quad e = (-11, 17)$, $f = (1, 45)$ gives $s = 2$, $f \rightarrow (23, 11)$.

(iii) $\quad e = (-11, 17)$, $f = (23, 11)$ so $s = 0$

Hence, $l_1 = 0.0395$, $l_2 = 0.0498$, $v_2 = 20.25$, $r = 1.26$. $\qquad\square$

In three or more dimensions we can give bounds on the lattice constants from Theorem 2.11. The vectors e_j^* defined there are known as the *dual basis*. They are the rows of E^{-1}, where E has columns $e_1 \cdots e_k$.

Theorem 2.11. Let (e_i) be any basis of Λ_k with increasing $\|e_i\|$. Let $e_1^*, \ldots, e_k^*$ be defined by $e_i^T e_j^* = \delta_{ij}$, and $w = \min \|e_k^*\|$. Then

(i) $\quad 1/w, \dfrac{1}{N}\displaystyle\prod_1^{k-1} \|e_i\| \leqslant l_k \leqslant \|e_k\|$

(ii) $\quad 1/\|e_k\| \leqslant v_k \leqslant w$

PROOF.   From the definitions and Theorem 2.16.   □

To use Theorem 2.11 we need to find a basis made up of short vectors. We can extend the ideas of Theorem 2.10 by:

Algorithm 2.3.   Fix some order of the pairs $\{i, j\}$. Apply the following until no change is made for any pair:
   Assume $\|e_i\| \leqslant \|e_j\|$. Let $s = \text{nint}(e_i^T e_j / \|e_i\|^2)$. If $s \neq 0$, replace e_j by $e_j - se_i$.

Exactly as for $k = 2$ this will terminate in a finite number of steps and find a basis with shorter vectors. In most cases the right-hand inequalities in Theorem 2.11 are then equalities, but not always. The bounds are usually quite close.

Example. $k = 3, N = 2^{16}, a = 249$. Using $\{1, 2\}, \{1, 3\}, \{2, 3\}$ as the order reduces the basis (2.3) to

$$Ne_1 = (260, -796, -1596)$$

$$Ne_2 = (-519, 1841, -343)$$

$$Ne_3 = (1316, 4, 996)$$

and $e_1 + e_2 + e_3$ is shorter than e_2, the longest of these vectors, but even so Theorem 2.11 yields $0.0259 \leqslant l_3 \leqslant 0.0296$.   □

This suggests trying further transformations, as does

Theorem 2.12. (Minkowski).   For $k = 3$ or 4, basis e_i has $\|e_i\| = l_i$, $i = 1, \ldots, k$ provided
   (i) $\|e_1\| \leqslant \cdots \leqslant \|e_k\|$ and
   (ii) $\|e_i\| \leqslant \|e_i + \sum_{j < i} c_j e_j\|$, each $c_j \in \{0, +1, -1\}$.

PROOF.   Section 2.7.   □

In applying Algorithm 2.3 we have already checked all combinations with just one $c_i \neq 0$. ($s = 0$ implies $\|e_i \pm e_j\| \geqslant \|e_i\|$.) For $k = 3$ this leaves four combinations for e_3, and for $k = 4$ four for e_3 and 20 for e_4. If we do find a shorter vector, we can replace e_i by that vector and repeat Algorithm 2.3 and the test of Theorem 2.12. In our example this gives $l_3 = 0.0275$.

This gives us a way to find l_1, l_k, and r exactly for $k \leqslant 4$. For $k \geqslant 5$ it is possible that no basis attains $l_1, \ldots, l_k$, but the bounds of Theorem 2.11 almost always suffice after applying Algorithm 2.3.

The Spectral Test

The vectors (e_i^*) introduced in Theorem 2.11 form a basis for another lattice Λ^*, known as the *polar* (or dual) lattice.

Theorem 2.13. Let $\mathcal{H} = \{x|x^T u \text{ integer}\}$ be a family of parallel hyperplanes covering Λ. Then $u \in \Lambda^*$ and $v_k = l_1^*$, the length of the shortest nonzero vector in Λ^*.

PROOF. The basis (e_i^*) spans $\mathbb{R}^k$, so $u = t_1 e_1^* + \cdots + t_k e_k^*$ for *real* t_i. For each i, $e_i \in \mathcal{H}$, so $t_i = e_i^T u$ is an integer; hence $u \in \Lambda^*$. Conversely, if $u \in \Lambda^*$, $u \neq 0$, then all $e_i \in \mathcal{H}$ and so $\Lambda \subset \mathcal{H}$.

The spacing of $\mathcal{H}$ is $\min\{\|x\| \mid x^T u = 1/\|u\|\}$, so $v_k = \min\{\|u\| \mid \mathcal{H} \supset \Lambda\} = \min\{\|u\| \mid 0 \neq u \in \Lambda^*\} = l_1^*$. $\square$

We can rewrite this conclusion by noting that $u \in \Lambda^*$ if and only if $e_1^T u \cdots e_k^T u$ are integers. Thus

$$v_k = \min\{\|u\| \mid 0 \neq u, u_i \text{ integers}, u_1 + au_2 + \cdots + a^{k-1}u_k \text{ is a multiple of } N\}$$

Such a quantity was defined by Coveyou and Macpherson (1967), who called testing for large values of v_k the "spectral test." In that context "large" is often assessed by forming $\mu_k = \omega_k v_k^k/N$, where $\omega_k = \pi^{k/2}/\Gamma(k/2 + 1)$ is the volume of the unit ball in $\mathbb{R}^k$. Values of μ_k larger than 1 are thought good. We can use Theorem 2.16 on Λ^* to show that $v_k \leqslant c_k N^{1/k}$, so $\mu_k \leqslant \omega_k c_k^k$, which helps explain why values greater than 1 are thought good. It seems preferable to use the inequality for v_k in a similar way to $r \geqslant 1$, remembering that v_k is an absolute measure of the granularity of Λ_k.

It remains to find v_k. For v_2 we have Theorem 2.10. For $k = 3$ or 4 we could apply Algorithm 2.3 and Theorem 2.12 to the polar lattice Λ^*. However, if the bounds of Theorem 2.11 are not sufficient, we can carry out a finite search by

Theorem 2.14 (Dieter, 1975). $v_k = \min\{\|u\| \mid u = t_1 e_1^* + \cdots + t_k e_k^* \neq 0,$ $|t_i| \leqslant \text{int}[w\|e_i\|]\}$ for any polar basis (e_i^*).

PROOF. $|t_i| = |e_i^T u| \leqslant \|e_i\| \|u\|$ by Cauchy-Schwartz. For the minimal u, $\|u\| = v_k \leqslant w$ by Theorem 2.11. $\square$

In practice $\text{int}[w\|e_i\|]$ is almost always 0 or 1, and by Theorem 2.12 we can take it to be 1 for $k \leqslant 4$. One further useful trick [from Knuth (1981)] is to take as the starting basis for Λ_k the basis (2.3) transformed by the transformations used for Λ_{k-1}, and to update the polar basis (e_i^*) with (e_i). Fortran code is given by Hopkins (1983) and in Appendix B.

Example. $N = 512, a = 45$ (continued).   The final basis for Λ_2 was

$$Ne_1 = (-11, 17), \qquad Ne_2 = (23, 11)$$

This gives us as initial basis

$$Ne_1 = (-11, 17, 253)$$
$$Ne_2 = (23, 11, -529)$$
$$Ne_3 = (0, 0, 512)$$

since $a^2 \equiv -23 \bmod N$, and the last element must be a^2 times the first. From this we have

$$e_1^* = (-11, 23, 0)$$
$$e_2^* = (17, 11, 0)$$
$$e_3^* = (23, 0, 1) \qquad (23 \equiv -a^2)$$

We now apply Algorithm 2.3, noting that if $e_j \to e_j - se_i$, then $e_i^* \to e_i^* + se_j^*$, to yield

$$Ne_1 = (23, 11, -17) \qquad e_1^* = (12, 6, -10)$$
$$Ne_2 = (22, -34, 6) \qquad e_2^* = (7, -10, 3)$$
$$Ne_3 = (82, 106, 162) \qquad e_3^* = (1, 1, 2)$$

when reordered in increasing length. This passes Theorem 2.12, so $l_1 = 0.0598$, $l_3 = 0.411$, $r = 6.86$, and $w = \sqrt{6}$. We have $|t_1| \leqslant 0$, $|t_2| \leqslant 0$, $|t_3| \leqslant 1$ in Theorem 2.14, so $v_3 = w$ attained at e_3^*. Note that $v_3 l_3 = 1.007$.   ☐

Assessing Congruential Generators

Table 2.4 lists some of the results of applying the preceding methods to $\Lambda_2, \Lambda_3, \Lambda_4$ for some commonly used generators. Only r and v_k are shown, since in all cases l_k is very close to $1/v_k$.

Line 1 is the generator G05CAF of the NAG Fortran library. Line 2 from Marsaglia (1972) is used by DEC for its VAX compilers. Line 7 is from CDC Fortran (FTN 4.x and 5.x compilers). All seem quite acceptable. Line 3 is used by BASIC on the Sinclair ZX81 (Tootill, 1982). Figure 2.3 confirms its two-dimensional granularity, which is due to both a bad choice of multiplier (Exercise 2.13) and too short a period. Lines 4 and 5 are for IBM 360/370

Table 2.4. Lattice Criteria for Certain Congruential Generators

	M	a	c	$k=2$		$k=3$		$k=4$	
				r	v	r	v	r	v
1	2^{59}	13^{13}	0	1.23	3.44×10^8	1.57	4.29×10^5	1.93	1.55×10^4
2	2^{32}	69069	1	1.06	6.51×10^4	1.29	1440	1.30	230
3	$2^{16}+1$	75	0	11.7	75	1.59	31.4	3.43	9.17
4	$2^{31}-1$	7^5	0	7.60	1.68×10^4	3.39	639	2.07	147
5	$2^{31}-1$	630360016	0	1.29	4.09×10^4	2.92	625	1.64	201
6	2^{35}	8404997	1	2.81	1.11×10^5	1.93	2930	5.98	147
7	2^{48}	44, 485, 709, 377, 909	0	1.29	7.45×10^6	1.85	3.44×10^4	3.85	1370
8	2^{32}	2147001325	715136305	1.13	6.40×10^4	1.09	1540	1.16	269
9	10^8+1	23	0	1.89×10^5	23	8211	23	357	23
10	10^9	314159221	211324863	3.89	1.61×10^4	2.12	800	2.46	103
11	2^{48}	5^{17}	1	1.87	1.23×10^7	2.86	4.74×10^4	1.67	3400
12	$2^{31}-1$	397204094	0	2.82	2.77×10^4	2.63	832	1.50	171

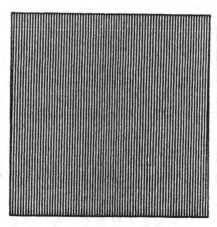

Figure 2.3. Plot of all pairs (U_i, U_{i+1}) from the Sinclair ZX81 generator, $X_i = 75X_{i-1}$ mod$(2^{16} + 1)$.

series machines from Lewis et al. (1969) and Payne et al. (1969). Line 6 is the default generator of GLIM3 (Baker and Nelder, 1978), a widely used statistical package, and line 8 is that built into BCPL (Richards and Whitby-Strevens 1979), a progenitor of C. Line 9 is Lehmer's original suggestion and lines 10 and 11 are from van Es et al. (1983). Lines 4 and 12 are routines GGUBFS and GGUBT of IMSL. Many other generators have been tested in this way, and the calculations can even be done on a microcomputer (Ripley, 1983b).

*Theory

Let E be a $k \times k$ matrix whose columns form a basis for a lattice Λ. Let $d(\Lambda)$ be the modulus of the determinant of E. Then $d(\Lambda)$ is the k-dimensional volume of the basic lattice "cell."

Theorem 2.15.

$$d(\Lambda) \leqslant \prod_1^k l_i \leqslant (c_k)^k d(\Lambda)$$

where

$$k = 2 \quad 3 \quad 4 \quad 5 \quad 6 \quad 7 \quad 6$$
$$(c_k)^{2k} = \tfrac{4}{3} \quad 2 \quad 4 \quad 8 \quad \tfrac{64}{3} \quad 64 \quad 256$$

PROOF.   Cassels (1959, Appendix; 1978, Section 12.2).   □

For our congruential generators $d(\Lambda_k) = 1/N$ from (2.3).

Theorem 2.16.

$$1 \leqslant v_k l_k \leqslant \prod_1^k l_i / d(\Lambda) \leqslant (c_k)^k$$

PROOF.   (i) Choose linearly independent vectors $\mathbf{f}_1, \ldots, \mathbf{f}_k \in \Lambda_k$ with $\|\mathbf{f}_i\| = l_i$, and $\mathbf{g} \in \Lambda_k^*$ with $\|\mathbf{g}\| = v_k = l_1^*$. Since $\{\mathbf{f}_i\}$ span $\mathbf{R}^k$, $\mathbf{g}^T \mathbf{f}_s \neq 0$ for at least one s. Now $\mathbf{g} = \sum t_i \mathbf{e}_i^*$, $\mathbf{f}_s = \sum s_i \mathbf{e}_i$, so $\mathbf{g}^T \mathbf{f}_s = \sum s_i t_i$ is an integer and

$$1 \leqslant |\mathbf{g}^T \mathbf{f}_s| \leqslant \|\mathbf{g}\| \, \|\mathbf{f}_s\| = v_k l_s \leqslant v_k l_k$$

by Cauchy–Schwartz.

(ii) Let Λ_{k-1} be the lattice with basis $\mathbf{f}_1, \ldots, \mathbf{f}_{k-1}$ and $H = \text{span}(\Lambda_{k-1})$. Consider parallel hyperplanes to H covering Λ_k with maximal spacing S. Let $\mathbf{f}$ be any member of Λ_k on a nearest hyperplane to the origin (excluding H). Let $\Lambda^\dagger$ be the lattice with basis $(\mathbf{f}_1, \ldots, \mathbf{f}_{k-1}, \mathbf{f})$. Then $\Lambda^\dagger \subset \Lambda_k$ and so $d(\Lambda_k) \leqslant d(\Lambda^\dagger) = S d(\Lambda_{k-1})$ by volume = basal area × height. Thus $S \geqslant 1/N d(\Lambda_{k-1})$, so

$$v_k \leqslant 1/S \leqslant N d(\Lambda_{k-1}) \leqslant N \prod_1^{k-1} l_i$$

and $v_k l_k \leqslant N \prod l_i \leqslant (c_k)^k$ by Theorem 2.15.   .   □

This result implies that $1/v_k$ and l_k are essentially equivalent measures of "granularity" of a lattice. The relation with the ratio $r = l_k/l_1$ comes from:

Theorem 2.17.

(i) For $k = 2$, $\sqrt{(r/N)} \leqslant l_2 \leqslant c_2 \sqrt{(r/N)}$

   $\sqrt{(N/r)} \leqslant v_2 \leqslant c_2 \sqrt{(N/r)}$

(ii) For $k \geqslant 2$, $1 \leqslant r \leqslant N l_k^k$

   $N^{-1/k} \leqslant l_k \leqslant c_k N^{-1/k} r^{(1 - 1/k)}$

PROOF.   (i) From Theorem 2.15 $1/N \leqslant l_1 l_2 = l_2^2/r = r v_2^2/N \leqslant c_2^2/N$

(ii) $1/N \leqslant l_1 \cdots l_k \leqslant l_k^k$, so $l_k \geqslant N^{-1/k}$

   $c_k^k/N \geqslant l_1 \cdots l_k \geqslant l_k l_1^{k-1} = l_k^k/r^{k-1}$

   $1/N \leqslant l_1 l_k^{k-1}$, so $r = l_k/l_1 \leqslant l_k^k N$ □

2.5. SHUFFLING AND TESTING

The best we have been able to do in the theoretical analysis of Sections 2.2–2.4 is to find for a limited class of generators the exact distribution of k-tuples $(U_i, \ldots, U_{i+k-1})$ over a period. Even this says nothing about the distribution of k-tuples over less than a period. For example, Fushimi and Tezuka (1983) tested pairs from the GFSR of period $2^{521} - 1$ considered by Bright and Enison (1979); although this *is* 2-distributed, the part of the period tested of length about 10,000 had considerably too many pairs near the diagonal of the unit square. We again discover philosophical problems, for such events will happen with true random numbers, and by asking for our pseudo-random numbers to conform too closely to expectation we will damage their credibility for some purposes. It does seem essential to test several subsequences from a generator with different starting points before jumping to conclusions.

There is a mistaken belief that taking seeds widely spaced apart in (X_i) and running the same congruential generator with these seeds will give "more independent" streams than sampling from a single sequence. Consider seeds X_0 and $Y_0 = X_j$. Then $Y_i = X_{i+j} = \{a^j X_i + (a^j - 1)c/(a - 1)\}$ mod M, so $\{(X_i, Y_i)\}$ lie on a lattice corresponding to the multiplier $(a^j \bmod M)$. It is entirely possible that $\{(X_i, Y_i)\}$ has much coarser structure than $\{(X_{2i}, X_{2i+1})\}$ and extremely unlikely that it has better lattice constants. If (X_i) is not thought sufficiently random to be used as the sole source of random numbers, one needs a better generator!

Shuffling

Various methods are available to modify the output of a suspect generator. They are not recommended since they are little understood, but they may provide a quick "fix" where necessary.

A. Generate output in blocks of length L, and apply a fixed permutation to each block before use. This should be sufficient to repair RANDU, for example. [See Atkinson (1980).]

B. Apply a random shuffle to (U_i). Suppose we have $T[0], \ldots,$ $T[k-1]$ initially filled with $U_1, \ldots, U_k$, and a second pseudo-random sequence (V_i). At each step we use V_n to select a random member of T; that is, we set $J = \text{int}[kV_n]$, then return $T[J]$ and replace it by U_n. This idea is due to MacLaren and Marsaglia (1965).

C. A subtly different method to B was proposed by Bays and Durham (1976). In their method the last value output is used rather than V_n to choose the next member of T to be replaced. Whereas B can make a sequence worse if (U_i) and (V_i) are closely related, no such examples are known for the Bays–Durham method. (They may of course exist.)

D. Given two sequences (X_i) and (Y_i) with moduli M, set $U_i = (X_i + Y_i)/M$ mod 1. An extension of this idea to three sequences was used by Wichmann and Hill (1982). Determining the period can be tricky, as those authors found.

E. Instead of adding we could form $U_i = (X_i \, \text{EOR} \, Y_i)/M$.

The only theoretical analysis of these schemes have been on simplified versions that may not be reliable models—Bays and Durham (1976), Brown and Solomon (1979), Nance and Overstreet (1978), and Rosenblatt (1975).

Empirical Testing

Any significance test of independence or uniformity or both can be applied to the output $(U_1, \ldots, U_n)$ of a pseudo-random number generator. Many tests have been used and it is most convenient to group them according to the property tested.

Tests for Independence

Any nonparametric test for independence can be applied to (U_i) or $(Y_i = \text{int}[U_i \times K])$ for any integer K. Often it is easiest to take K a power of 2 and so examine the first few bits of X_i.

(a) *Gaps Test for* (U_i). Fix constants $0 < \alpha < \beta < 1$ and consider the lengths of intervals for which $U_i \notin (\alpha, \beta)$. If the sequence (U_i) is independent, the distribution of lengths should be geometric with parameter $P(\alpha < U_1 < \beta) = (\beta - \alpha)$. Furthermore, independence means that successive gap lengths are independent, so we can compare observed and empirical distributions by a chi-squared test. As an example, consider Table 2.1 with $\alpha = 0.4, \beta = 0.6$. Then the gap lengths are 0, 7, 1, 0, 1, 0, 8, 1, 5, 1, 6, 7 so

$k =$	0	1	2	3	4	5	6	7	8	>8
Observed	3	4	0	0	0	1	1	2	1	0
Expected	2.4	1.92	1.54	1.23	0.98	0.79	0.63	0.50	0.40	1.61

which needs no statistical test to reject independence.

(b) *Runs Test for* (U_i). Runs up are monotone increasing subsequences; runs down are defined by replacing increasing by decreasing. There are several subtly different "runs tests" depending on whether both runs up and down are used, and on the exact definition of a run. Probably the easiest is to discard the first element of a run, so the first two runs up in Table 2.1 are (0.563, 0.624) and (0.811, 0.999). In that case the run-up lengths are independent and a chi-squared test can be used to compare their observed and expected frequencies. Tedious but elementary probability shows that E(number with run length $= k) = (n + 1)k/(k + 1)! - (k - 1)/k!$, $k = 1, \ldots, n$ for a sequence of length n. Barton and Mallows (1965) discuss runs tests in more detail.

(c) *Permutation Tests for* (U_i). Divide (U_i) into blocks of length t, $(U_1, \ldots, U_t)$, $(U_{t+1}, \ldots, U_{2t})$, ... There are $t!$ possible orderings of a block of t distinct numbers, and these should be equally probable. Counting the occurrences of all possible orderings and using a chi-squared test gives us a test for independence. This is only useful for moderate t, since we will need $n \geqslant t!$ We saw in Section 2.1 that the Fibonacci sequence (1.2) will fail this test for $t = 3$.

(d) *Coupon Collectors' Test for* (Y_i). Consider the lengths of sequences needed to "collect" all integers $0, \ldots, K - 1$. This gives us a frequency distribution on $\{K, K - 1, \ldots\}$. The probability of a length r being needed is found by combinatorial arguments (Greenwood, 1955).

Note that (a) and (d) depend on uniformity, whereas (b) and (c) work for any continuous marginal distribution of the (U_i).

Tests for Uniformity

Tests for uniformity can be any nonparametric test of a known distribution. The most commonly used are a chi-squared test based on dividing (0, 1) into intervals, and the Kolmogorov–Smirnov test $\max|F_n(x) - x|$, where $F_n(x) =$ (number of $U_i \leqslant x$)/n, the empirical distribution function of $(U_1, \ldots, U_n)$. Its computation is discussed by Gonzalez et al. (1977).

Tests of Pairs, and k-tuples

The chi-squared test can also be applied to test the uniformity of k-tuples $\{(U_{ki}, \ldots, U_{ki+k-1})\}$, dividing $[0, 1]^k$ into a number of small regions. To do so effectively and ensure a reasonable number of counts in each cell of the test needs a very large number of observations, so this tends to be a weak test. Note that this test cannot be applied to $\{(U_i, \ldots, U_{i+k-1})\}$ since these k-tuples are not independent. Good (1953, 1957) provides a correct modified test for pairs $\{(U_i, U_{i+1})\}$.

An alternative is to use time-series methods to examine the correlation structure of (U_i). These methods are meant for normally distributed sequences, and it may be better to apply them to $V_i = \Phi^{-1}(U_i)$, which is normally distributed if Φ is the cumulative distribution function for the normal (see Theorem 3.1). We can then test whether the correlation between U_i and U_{i+t}, or V_i and V_{i+t}, is zero. This is again a weak test, for lack of correlation does not imply independence.

More sensitive tests are provided by tests of the k-tuples as a point pattern. Theorem 2.6 provides one such test statistic (Ripley and Silverman, 1978) and others are described in Ripley (1981, Chapters 7 and 8).

The theoretical tests of Sections 2.2–2.4 have been found to be more powerful than empirical tests in the sense that "good" generators by the theoretical criteria have been found to fail the empirical tests no more often than would be expected by chance. Nevertheless it is always worth conducting some empirical tests to check that the generator has been implemented correctly. (Microcomputer implementations work incorrectly surprisingly often from faulty compilers or side effects of operating systems.)

We would of course expect the statistical tests to be failed occasionally by chance. In extensive investigations it is a good idea to try each test on a large number of nonoverlapping subsequences of (U_i). For each *test* we obtain an observation of either pass/fail or a significance level and can test these observations against their known distribution. Perhaps the most commonly used example of this procedure is to apply the Kolmogorov–Smirnov test of uniformity. This gives rise to a significance level P uniformly distributed on $(0, 1)$, and the Kolmogorov–Smirnov test is applied again to the observed significance levels. An example is given by van Es et al. (1983).

2.6. CONCLUSIONS

The net effect of both theoretical analysis and empirical investigations is that a good pseudo-random number generator should:

(a) use a simple algorithm and so be rapid, taking considerably less time than evaluating a logarithm;

(b) be periodic with a long period, at least 2^{27} or 10^8, and take values evenly spread in $[0, 1)$, preferably excluding zero;

(c) have k-tuples for $k = 2, 3, 4$ and preferably $k \leq 10$ as uniformly distributed as possible in $[0, 1)^k$;

(d) have been checked carefully to see that it does implement the stated algorithm.

Unfortunately a large proportion of generators in common use fail to have some of these properties, principally b and c. However, a number of generators are available with the desired properties which are fairly simple to implement.

Among congruential generators, line 2 of Table 2.4,

$$X_i = (69069X_{i-1} + 1)\bmod 2^{32}, U_i = 2^{-32} X_i$$

has been implemented successfully in Fortran and assembler on a range of machines from 8-bit microcomputers to 64-bit supercomputers. If the NAG library is available, line 1 is an obvious choice. It is preferable to have a generator with period greater than 2^{32}, and it will often be possible to use line 11 with period 2^{48}. Implementing these generators in a high-level language is normally done using double-precision reals, which usually can represent exactly considerably larger integers than integer types. (See Appendix B.1.)

The GFSR generators represent an easier solution in environments with only limited precision arithmetic, provided a word-wise EOR operation is available. Generally we will choose at least 15-bit integers and ask for at least 4-distribution. One such recommendation is based on $p = 98$, $q = 27$, $L = 15$, which has a "granularity" of 2^{-15} and almost exact independence in up to six dimensions. To initialize it we take $b_i = b_{i-98}$ EOR b_{i-71} and make up Y_i out of $(b_i, b_{i+6}, b_{i+12}, \ldots, b_{i+84})$, $i = 1, \ldots, 98$ and then use algorithm 2.2.

It is always helpful to have two or more generators available and to run important simulations using each, to reduce the likelihood that anomalous results are due to the quirks of the generator used.

*2.7. PROOFS

We need three lemmas for the results of Section 2.2.

Lemma A.   Let $p_1^{z_1} \cdots p_r^{z_r}$ be the prime factorization of M. Then the period of any congruential generator with modulus M is the lowest common multiple of its periods modulo $p_i^{z_i}$.

PROOF.   By induction we need only consider $M = m_1 m_2$ with $\gcd(m_1, m_2) = 1$. Let $Y_j = X_j \bmod m_1$, $Z_j = X_j \bmod m_2$, with X_0 any value in the periodic cycle. Suppose X_i, Y_i, Z_i have periods d, d_1, and d_2. Then $Y_j = Y_0$ iff j is a multiple of d_1. Since $X_d = X_0$, $Y_d = Y_0$, so d is a multiple of d_1. By symmetry it is a multiple of d_2 and hence of $\mathrm{lcm}(d_1, d_2) = l$. Now $X_l - X_0$ is a multiple of both m_1 and m_2, since $Y_l = Y_0$, $Z_l = Z_0$, hence of M. Thus $X_l = X_0$ and $d = l = \mathrm{lcm}(d_1, d_2)$.   □

Lemma B (Fermat, 1640). Suppose p is prime and $0 < a < p$. Then $a^{p-1} \equiv 1$ mod p.

PROOF. Consider $\{ra \bmod p | 0 \leqslant r < p\}$. This is a set of p distinct numbers ($ra \equiv sa$ mod p implies $r = s$), so it is $\{0, \ldots, p-1\}$. Thus $\{a \bmod p, 2a \bmod p, \ldots, (p-1)a \bmod p\} = \{1, \ldots, p-1\}$ and

$$\prod_1^{p-1} ra \equiv \prod_1^{p-1} r \bmod p$$

whence $a^{p-1} \equiv 1$ mod p. $\qquad\square$

Lemma C. (X_{ki}) has period $d/\gcd(k, d)$ if (X_i) has period d.

PROOF. $X_{ki} = X_0$ if and only if ki is a multiple of d if and only if i is a multiple of $d/\gcd(k, d)$. $\qquad\square$

Proof of Theorem 2.3.

(i) Since $c = 0$, $X_i = a^i X_0$ mod M. Let $p_1^{\alpha_1} \cdots p_r^{\alpha_r}$ be the prime factorization of M. Then $(X_j \bmod p_i^{\alpha_i})$ has period at most $(p_i^{\alpha_i} - 1)$ (omitting zero), so (X_j) has period at most $\prod_1^r (p_i^{\alpha_i} - 1)$, $< M - 1$ unless $r = 1$.

(ii) Suppose $M = p^\alpha$ for a prime p, $\alpha > 1$. Then $(X_i \bmod p)$ has period d dividing $(p - 1)$ by lemma B. Thus $(X_{id} - X_0)$ is a multiple of p, whence (X_{id}) has period at most $p^{\alpha-1}$, and (X_i) has period at most $dp^{\alpha-1} < M - 1$.

(iii) Suppose M is prime. By lemma B the period d divides $M - 1$. Suppose $M - 1 = nd$, and that p is a prime factor of n. Then $s = (M - 1)/p$ is a multiple of d, and $X_s = a^s \bmod M = X_0$, so $a^s \bmod M = 1$. Hence if $n > 1$, a is not a primitive root. Conversely, if a is not a primitive root, let $s = (M - 1)/p$, when $X_s = a^s X_0 \bmod M = X_0$, so the period divides s and is less than $M - 1$. $\qquad\square$

Proof of Theorem 2.4

Let $b = a^k \bmod M$. Then (X_{ki}) corresponds to multiplier b, and has period $(M - 1)$ if and only if $\gcd(k, M - 1) = 1$ by lemma C. $\qquad\square$

Proof of Theorem 2.1

(i) By lemma A we can confine attention to $M = p^\alpha$, p prime.

(ii) If $c = 0$, the period is at most $M - 1$ by the proof of Theorem 2.3. We assume $c > 0$.

(iii) If $a = 1$, $X_i = (X_0 + ic) \bmod M$, so $X_i = X_0$ iff $ic \bmod M = 0$ iff i is a multiple of $M/\gcd(c, M)$, the period. We can now assume $M = p^\alpha$, $c > 0$, $a > 1$. We have

$$X_{i+j} = \left\{ a^j X_i + \frac{(a^j - 1)c}{(a - 1)} \right\} \bmod M \text{ for } j \geqslant 0$$

(iv) Suppose the period is M. Then we may take $X_0 = 0$. Some $X_i = 1$, when $(a^i - 1)c/(a - 1) \equiv 1 \bmod M$ and hence $\gcd(c, M) = 1$ [for $(a^i - 1)/(a - 1) = 1 + a + \cdots + a^{i-1}$, an integer]. Also, $X_M = 0$, so $(a^M - 1)c/(a - 1)$ is a multiple of $M = p^\alpha$. If $a \not\equiv 1 \bmod p$ this implies $a^M - 1 \equiv 0 \bmod M$, hence $a^M \bmod p = 1$. However, lemma B shows $a^p \equiv a \bmod p$, so $a^M \equiv a \bmod p$. We conclude $a \equiv 1 \bmod p$.

(v) Suppose $M = 2^\alpha$, $a \geqslant 2$. If $a \equiv 1 \bmod 2$ but $a \not\equiv 1 \bmod 4$, $a \equiv 3 \bmod 4$. Then

$$X_i = \{a^2 X_{i-2} + (a + 1)c\} \bmod M$$

hence $X_2, X_4, \ldots$ are multiples of $(a + 1)c \bmod M = 4c \bmod M$. Thus (X_{2i}) takes at most $M/4$ values, and (X_i) has period at most $M/2$.

This establishes necessity. For sufficiency assume $M = p^\alpha$. We will use induction on α. For $\alpha = 1$, $a \equiv 1 \bmod M$, whence $(X_i) = (ic \bmod M)$ starting from zero, which has period M. Now suppose the theorem holds for $M = p^{\alpha-1}$. Fix $X_0 = 0$. From the conditions $a = 1 + qp^e$ for $p^e > 2$. Thus

$$a^p = (1 + qp^e)^p = 1 + pqp^e + \cdots + q^p p^{ep} = 1 + sqp^{e+1}$$

for an integer s with $s \equiv 1 \bmod p$. Now $X_p = (a^p - 1)c/(a - 1) \bmod p^\alpha = sqp^{e+1}c/qp^e \bmod p^\alpha = scp \bmod p^\alpha$. By induction we find X_{ip} is a multiple of p for all i. Let $Y_i = X_{ip}/p$. Then

$$Y_i = \{a^p Y_{i-1} + (a^p - 1)c/(a - 1)p\} \bmod p^{\alpha-1}$$
$$= \{a^p Y_{i-1} + sc\} \bmod p^{\alpha-1}$$

Applying the theorem for modulus $p^{\alpha-1}$ shows Y_i has period $p^{\alpha-1}$ since $s \equiv 1 \bmod p$, so $\gcd(sc, p^{\alpha-1}) = 1$. We can deduce $X_M = 0$, so (X_i) has period dividing $M = p^\alpha$. However $X_{p^\alpha-1} = pY_{p^\alpha-2} \neq 0$, and (X_i) has period M. $\square$

Proof of Theorem 2.2

(i) Suppose X_0 is even. Then $X_0 = 2^r Y$, for Y odd, and

$$(X_i 2^{-r}) = a(X_{i-1} 2^{-r}) \bmod 2^{\beta-r}$$

reduces to the case of an odd seed.

(ii) Suppose a is even. Then $X_\beta = a^\beta X_0 \bmod 2^\beta = 0$. We now suppose X_0 and a are odd, so X_i is always odd. Choose X_0 as the smallest value in the period. Then

$$(X_i - X_0) = \{a(X_{i-1} - X_0) + (a - 1)X_0\} \bmod 2^\beta$$

and $(a - 1)$ is even, so $X_i - X_0$ is even, $= 2Y_i$ say.

$$Y_i = \{aY_{i-1} + a'X_0\} \bmod 2^{\beta-1}$$

where $a' = (a - 1)/2$. From Theorem 2.1 this has period less than $M/2$, for either a' is even or $a \not\equiv 1 \bmod 4$.

(iii) Suppose $a \equiv 1 \bmod 4$, so $a = 1 + 4b$. Then $a' = 2b$ and Y_i is even. Let $Z_i = Y_i/2$, with

$$Z_i = \{aZ_{i-1} + bX_0\} \bmod 2^{\beta-2}$$

From Theorem 2.1 this has period $M/4$ if b is odd ($a \equiv 5 \bmod 8$) otherwise less than $M/4$.

(iv) Suppose $a = 3 + 4b$. Then

$$Y_i = \{a^2 Y_{i-2} + a'(a - 1)\} \bmod 2^{\beta-1}$$

and $a'(a + 1) = 4(b + 1)(2b + 1)$ is a multiple of 4. Hence so is Y_{2i}. Let $W_i = Y_{2i}/4$, with

$$W_i = \{a^2 W_{i-1} + (b + 1)(2b + 1)\} \bmod 2^{\beta-3}$$

Now $a^2 \bmod 4 = (16b^2 + 24b + 9) \bmod 4 = 1$, so (W_i) has period $M/8$ only if b is even. Thus if $a \equiv 7 \bmod 8$, (Y_i) has period less than $M/8$. If $a \equiv 3 \bmod 8$, (Y_{2i}) are multiples of 4 with period $M/8$, whereas (Y_{2i+1}) are odd. Thus (Y_i) and (X_i) have period $M/4$.

(v) Suppose $a \equiv 5 \bmod 8$. Then $X_i = X_0 + 4Z_i$, so the smallest value in the sequence, $b = (X_i \bmod 4)$ for any i. Then $U_i = X_i/M = b/M + \{Z_i/(M/4)\}$ as required. $\qquad\qquad\square$

Proof of Theorem 2.5

Let $\oplus$ denote addition modulo M.

(i) $\{U_i\} = \{(X_0 \oplus s)/M \mid s = 0, 1, \ldots, M - 1\}$, so
$$\{(U_i, \ldots, U_{i+k-1})\} = \{(X_0 \oplus s, a(X_0 + s) \oplus c, \ldots)/M \mid s = 0, \ldots, M - 1\}$$
$$= [0, 1)^k \cap \{(X_0 + s, aX_0 + as + c + t_2M, \ldots)/M \mid s, t_2, \ldots, t_k \text{ integer}\}$$
$$= [0, 1)^k \cap \{(X_0, \ldots, X_{k-1})/M + \Lambda_k)$$

(ii) $\{(U_i, \ldots, U_{i+k-1})\} = \{(s, as, a^2s, \ldots)/M \mid s = 1, \ldots, M - 1\}$
　　$= (0, 1)^k \cap \{(s, as + t_2 M, \ldots)/M \mid s, t_2, \ldots, t_k \text{ integer}\}$
　　$= (0, 1)^k \cap \Lambda_k$

For the nonoverlapping k-tuples we note

$$\{U_{ki}\} = \{(U_0 + s/N) \bmod 1\}$$

by lemma C in the cases claimed and modify the above accordingly.

Proof of Theorem 2.12

(i) Suppose $g \in \Lambda_k$, $g = t_1 e_1 + \cdots + t_k e_k$. If $t_j \neq 0$ we will show $\|g\| \geqslant \|e_j\|$. This establishes $\|e_i\| = l_i$ for $i = 1, \ldots, k$.

(ii) By changing e_i to $-e_i$ if necessary we may assume all $t_i \geqslant 0$. Let $T = \max t_i$. We proceed by induction on T. If $T = 1$ then $\|g\| \geqslant \|e_j\|$ by hypothesis (ii). Suppose the result is true for $\max t_i \leqslant T - 1$. Let $m = \min\{t_i \mid t_i > 0\}$ and $r = \max\{i \mid t_i = m\}$. Define E as the sum of e_i over all indices i except r with $t_i > 0$. Let $h = g - mE$. We will show $\|g\| \geqslant \|h\|$. Now *either* $\max h_i \leqslant T - 1$ and $\|h\| \geqslant \|e_j\|$ *or* all nonzero t_i were equal to m, when $h = me_r$, so $\|h\| \geqslant \|e_j\|$. In either case $\|g\| \geqslant \|h\| \geqslant \|e_j\|$.

(iii) Fix $s < t$. Then $\|e_s + e_t\|^2 \geqslant \|e_s\|^2$, whence $2e_s^T e_t \geqslant - \|e_s\|^2$ and hence $e_s^T e_t \geqslant - \frac{1}{2}\|e_s\|^2$ whether $s < t$ or $s > t$. Now consider

$$(h - me_r)^T E = \sum_{i \neq r} h_i e_i^T E = \sum_{i \neq r \text{ in } E} h_i e_i^T E$$

$$= \sum h_i \{\|e_i\|^2 + (\text{up to } 2) e_i^T e_j\} \geqslant 0 \text{ since } h_i \geqslant 0$$

(iv) $\|g\|^2 - \|h\|^2 = \|h + mE\|^2 - \|h\|^2$

$$= 2m h^T E + m^2 \|E\|^2 = 2m(h - me_r)^T E + m^2 \{\|E + e_r\|^2 - \|e_r\|^2\}$$

which is nonnegative by (iii) and by hypothesis.　　　　　　　　　　□

EXERCISES

2.1.　Complete the sequence (1) and show that it eventually repeats. Try other starting values. Is the behavior starting from 8653 typical?

2.2.　Investigate all starting values for the two-digit decimal and eight-bit binary middle square methods.

2.3. Prove that for the *Fibonacci* recursion (2) that $U_{i-2} < U_i < U_{i-1}$ never occurs, whereas this event has probability $1/6$ for random numbers.

2.4. Try implementing the generator $U_i = 1013U_{i-1} \bmod 1$ in floating-point arithmetic and via (4) with $M = 10^5$ and $M = 2^{16}$. What periods are obtained?

2.5. Compute the outputs of the following congruential generators with $M = 64$. (a) $a = 29$, $c = 17$. (b) $a = 9$, $c = 1$. (c) $a = 13$, $c = 0$. (d) $a = 11$, $c = 0$.

2.6. Find the periods corresponding to multipliers 10, 12, 16, and 18 in a multiplicative congruential generator with $M = 67$.

2.7. Show for $M = r^\beta - s$, $Y = aX_{i-1} + c$, that $Y \bmod r^\beta + s(Y \operatorname{div} r^\beta) < (1 + s)M$.

2.8. Plot the lattices of (U_i, U_{i+1}) and (U_{2i}, U_{2i+1}) for the examples of Exercises 2.5 and 2.6.

2.9. Show that Algorithm 2.1 works.

2.10. Generate the shift-register sequence with $p = 7$, $q = 1$. Form the Tausworthe sequence with $t = L = 3$ and show that it is 2-distributed.

2.11. Generate the GFSR with $p = 7$, $q = 1$, starting 6, 5, 4, 3, 2, 1, 0 . . . . How does it fail to be 2-distributed?

2.12. Find starting values for the GFSR with $p = 7$, $q = 1$, $L = 3$ by the delay method and verify that it is 2-distributed both via Theorem 2.9 and by generating the sequence.

2.13. Find a better multiplier than 75 for $M = 2^{16} + 1$, $c = 0$.

2.14. Compute the lattice constants where appropriate for the examples of Exercises 2.5 and 2.6 in two and three dimensions.

2.15. Try the effect of the Bays–Durham shuffling algorithm on Table 2.1.

2.16. Apply the gaps and runs tests and the permutation test for $k = 3$ to both Table 2.1 and the output from Exercise 2.15.

2.17. Test empirically the pseudo-random number generators on all the computers you use. If possible, find out the algorithms claimed to be used, check that these are implemented as stated and pass appropriate theoretical tests. Alternatively, replace these generators with ones of known quality, and test your implementations.

CHAPTER 3

Random Variables

The scope of this chapter is the generation of independent random variables X_1, X_2, ... with a given distribution function F or probability density function (pdf) f. We assume that we have access to a supply (U_i) of random numbers, independent samples from the uniform distribution on $(0, 1)$. Our task is to transform (U_i) into (X_i). In most cases we will have the choice of several algorithms for doing so. Usually there will be no universal best choice; different methods might be recommended for once-off use and for adding to a computer center's mathematical library.

When choosing algorithms we will consider the following points.

(a) The method should be easy to understand and to program. It is all too easy to make mistakes while implementing sophisticated methods.

(b) The programs produced should be compact. This may only be important on small machines but can considerably reduce overheads in interpreted languages.

(c) The final code should execute reasonably rapidly. This point has been emphasized in the literature almost to the exclusion of the other two. Andrews (1976) cites a study in which generating the random variables cost 0.2% of the total computer usage of a simulation study. It is rarely important to save generation costs.

(d) The algorithms will be used with pseudo-random numbers and should not accentuate their deficiencies.

Experience has shown that the relative speeds of different algorithms vary surprisingly little across different computers and languages, APL being the main exception (Appleton, 1976). "Good" algorithms avoid large tables of constants and multiple calls to mathematical functions (ln, sin, cos, etc.)

54

RANDOM VARIABLES

3.1. SIMPLE EXAMPLES

Undoubtedly the best-known distribution is the normal distribution. One once popular way to sample from it is to use

$$X = \left(\sum_{1}^{12} U_i - 6 \right)$$

which has mean zero, variance one, and is approximately normally distributed by the central limit theorem. The approximation is fairly good, but curves of both pdf and cdf are discernibly different, with maximum differences of about 0.0050 and 0.0023 respectively (Exercise 3.1).

The best-known "exact" method for the normal distribution is that of Box and Muller (1958). (Note: *not* Müller as frequently given.)

Algorithm 3.1 (Box–Muller).

1. Generate U_1, set $\Theta = 2\pi U_1$.
2. Generate U_2, set $E = -\ln U_2$, $R = \sqrt{2E}$.
3. $X = R \cos \Theta$, $Y = R \sin \Theta$ are independent standard normal deviates.

For simplicity in programming often only X or Y is used. To understand how the algorithm works, consider a pair (X, Y) of standard normal deviates. Their joint pdf is

$$\frac{1}{2\pi} \exp[-\tfrac{1}{2}(x^2 + y^2)]$$

Let (R, Θ) be (X, Y) in polar coordinates. Then (R, Θ) has joint pdf

$$\left[\frac{1}{2\pi} \exp(-\tfrac{1}{2}r^2) \right] \left| \begin{matrix} \cos \theta & \sin \theta \\ -r \sin \theta & r \cos \theta \end{matrix} \right| = \left(\frac{1}{2\pi} \right) (re^{-r^2/2}) \text{ on } (0, \infty) \times (0, 2\pi)$$

when R and Θ are independent. Then $S = R^2 = X^2 + Y^2$ has a χ_2^2 distribution, which is also an exponential distribution of mean 2. Finally, let $U = \exp(-S/2)$. Then

$$P(U \leqslant u) = P(-S/2 \leqslant \ln u) = P(S \geqslant -2 \ln u)$$

$$= \exp[-\tfrac{1}{2}(-2 \ln u)] = u$$

for $0 \leqslant u \leqslant 1$. This transforms (X, Y) to (U, Θ), independent uniform random variables. Reversing the transformation yields the algorithm. (See Exercise 3.2.)

We have incidentally discovered a way to sample exponential variates.

Algorithm 3.2. $X = -\lambda^{-1} \ln U$ has an exponential (λ) distribution for
$$P(X \leqslant x) = P(U \geqslant \exp(-\lambda x)) = 1 - \exp(-\lambda x).$$

To sample from a Poisson distribution we can exploit the Poisson process. Let $E_i = -\ln U_i$ be the inter-event times in a Poisson process, and let N be the number of events by time t. Let $S_n = E_1 + \cdots + E_n, P_n = U_1 \times \cdots \times U_n$. Then $N = n$ if and only if $S_n \leqslant \mu < S_{n+1}$ if and only if $P_n \geqslant e^{-\mu} > P_{n+1}$. This gives the following.

Algorithm 3.3.

1. Set $P = 1, N = 0, c = e^{-\mu}$.
2. Repeat.
      Generate U_i, let $P = P \times U_i, N = N + 1$ until $P < c$.
3. $X = N - 1 \sim$ Poisson (μ).

We can now sample from χ^2 distributions, for χ^2_{2m} is the distribution of the sum of m independent exponential(2) variates, and χ^2_{2m+1} is the distribution of a χ^2_{2m} plus a squared standard normal. From normals and χ^2 we can obtain Student's t and the F distribution.

There are many other ingenious "tricks" of the type shown in these examples. The scope for invention is unlimited and hundreds of specific algorithms have been published. Most of them are based on a small number of general principles, the subject of the next section, with specific distributions being discussed in later sections.

*A Cautionary Tale

All our theory assumes the use of genuine random numbers. Neave (1973), Swick (1974), Chay et al. (1975), and Golder and Settle (1976) comment on the use of Algorithm 3.1 with congruential generators. Neave generated two million normal deviates from

$$X = \sqrt{-2 \ln U_i} \cos (2\pi U_{i+1})$$
$$Y = \sqrt{-2 \ln U_i} \sin (2\pi U_{i+1}) \tag{1}$$

which reverses the role of U_i and U_{i+1} in Algorithm 3.1. He reported that for the Y's all values lie in the range $(-3.3, 3.6)$ for the generator $a = 131$, $c = 0, M = 2^{35}$.

We know from Theorem 2.5 that many congruential generators have a lattice structure for $\{(U_i, U_{i+1})\}$, and (1) must transform this into some structure for $\{(X, Y)\}$. Figure 3.1 shows what can happen. In particular, Fig. 3.1d explains Neave's findings. His generator has $r = 2 \times 10^6$,

$l_2 = 0.0076$, $v_2 = 131$ and so has an appalling lattice structure. The points (X, Y) clearly lie on a spiral. We can see this from

$$X = \sqrt{-2 \ln U_i} \cos (2\pi a U_i + 2\pi c/M)$$
$$Y = \sqrt{-2 \ln U_i} \sin (2\pi a U_i + 2\pi c/M)$$

(by the periodicity of the trignometric functions), which shows (X, Y) lies on the spiral

$$(\sqrt{-2 \ln t} \cos(2\pi at + C), \sqrt{-2 \ln t} \sin(2\pi at + C)), \qquad t \in (0, 1)$$

The maximum of y values is attained when $2\pi at \approx \pi/2$ in the multiplicative case, so is $\sqrt{2 \ln 4a}$. This suggests we need a to be large, although Fig. 3.1c shows that this is not sufficient. Neave (1973) analyzed the distribution of Y on the assumption that $U_i \sim U(0, 1)$. However, U_i has a discrete distribution that radically alters the theory.

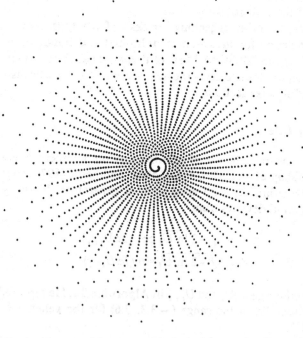

(a)

Figure 3.1. Plots of pairs (X, Y) from the Box–Muller algorithm 3.1 applied to real generators. (a) The congruential generator $X_i = (65X_{i-1} + 1) \bmod 2048$.

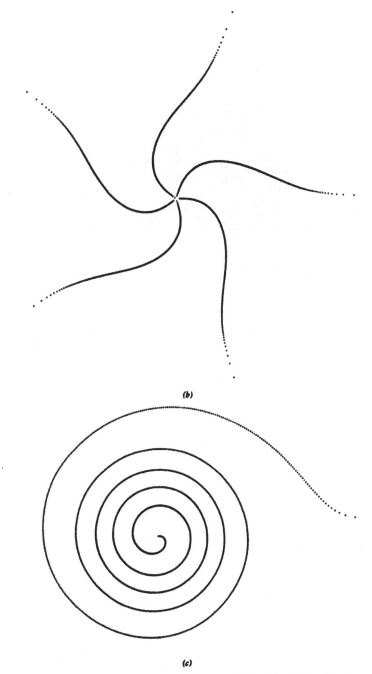

(b)

(c)

Figure 3.1. (*Continued*) (b) $X_i = (1229 X_{i-1} + 1) \bmod 2048$. (c) as (b), with U_i and U_{i-1} interchanged.

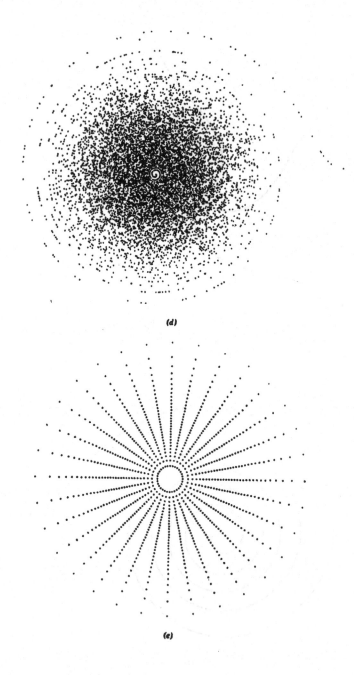

(d)

(e)

Figure 3.1. *(Continued)* (d) 10,000 points from Neave's example. (e) A GFSR with 2-distribution and $L = 5$.

Perhaps the best way to understand the Neave effect is to note that we can be concerned with large values of X and Y, hence with small values of U_i. The plots of $\{(U_i, U_{i+1})\}$ analyzed in Section 2.4 show which values of U_{i+1} occur with small U_i. It thus becomes possible to interpret the lattice behavior in terms of $\{(X, Y)\}$ plots.

Reversing the roles of U_i and U_{i+1} gives us an interpretation of the Algorithm 3.1 form of the Box–Muller transformation.

One might suppose that the near-independence of (R, Θ) from a GFSR illustrated in Fig. 2.2b might give a better distribution of X or Y than a congruential generator. Figure 3.1e shows that this is not necessarily so.

No general theory has yet been developed for the sensitivity of this or other algorithms to the random numbers used. Almost all the algorithms are continuous in the sense that nearby points in (U_i, U_{i+1}) space get mapped to nearby points in (X, Y) space, except perhaps for U_i or U_{i+1} near zero or one.

3.2. GENERAL PRINCIPLES

Almost all algorithms for sampling from specific distributions are derived from a few general principles. All the principles discussed in this chapter apply to continuous distributions. Most also apply to discrete distributions, but these are discussed in more detail in Section 3.3.

Inversion

It is well known in nonparametric statistics that if X has a continuous cdf F then $F(X) \sim U(0, 1)$. This suggests sampling from F by $X = F^{-1}(U)$ provided the inverse exists. The following theorem places *no* restrictions on F.

Theorem 3.1. Define F^- by $F^-(u) = \min\{x|F(x) \geq u\}$. Then if $U \sim U(0, 1)$, $X = F^-(U)$ is a sample from F.

PROOF. The minimum is attained by right-continuity of F, so $F(F^{-1}(u)) \geq u$, and $F^-(F(x)) = \min\{y \mid F(y) \geq F(x)\} \leq x$. Hence $\{(u, x) \mid F^-(u) \leq x\} = \{(u, x) \mid u \leq F(x)\}$ and $P(X \leq x) = P(F^-(U) \leq x) = P(U \leq F(x)) = F(x)$ as required. □

Examples. (a) Tossing a biased coin. Let $X = 1$ for heads, 0 for tails, $P(X = 1) = p$. Then $F(x) = 1 - p + pI(x \geq 1)$, so $F^-(u) = I(u \geq 1 - p)$. Thus $X = I(U \geq 1 - p) = I(1 - U \leq p) = I(U_1 \leq p)$, where $U_1 = 1 - U \sim U(0, 1)$.

(b) The exponential distribution has $F(x) = 1 - e^{-\lambda x}$ on $(0, \infty)$, so $F^-(U) = -\lambda^{-1} \ln(1 - U) = -\lambda^{-1} \ln U_1$, as in Algorithm 3.2.

(c) The Weibull distribution with $F(x) = 1 - \exp(-x^\beta)$ on $(0, \infty)$ has $X = (-\ln U_1)^{1/\beta}$

(d) The Cauchy distribution has pdf $f(x) = 1/\pi(1 + x^2)$ and $F(x) = \frac{1}{2} + \pi^{-1} \tan^{-1} x$, so $X = F^-(U) = \tan[\pi(U - \frac{1}{2})]$

(e) For the normal distribution we obtain $X = \mu + \sigma\Phi^{-1}(U)$. Now Φ^{-1} is just another transcendental function like ln or sin and can be approximated in the same way. Bailey (1981) gives some approximations, and Beasley and Springer (1977) give a Fortran program to evaluate Φ^{-1}.

Inversion is a universal method but may be too slow unless subprograms to calculate F^- are available. For example, it is theoretically possible to use inversion to sample from the beta density

$$f(x) = x^{\alpha-1}(1 - x)^{\beta-1}/B(\alpha, \beta) \quad \text{on} \quad (0, 1)$$

or the von Mises distribution

$$f(\theta) = \exp(\kappa \cos \theta)/2\pi I_0(\kappa) \quad \text{on} \quad (0, 2\pi)$$

but much simpler methods are given below.

Most of the pseudo-random number generators described in Chapter 2 give $\{U_i\}$ as a set of equally spaced points in $[0, 1)$. The inversion transformation will map this to a good approximation to the true distribution. However, the *independence* of $X_i = F^-(U_i)$ still depends on the independence of (U_i).

Rejection

Suppose we wish to sample from a pdf f but have a way to sample from the pdf g. Rejection methods (von Neumann, 1951)—sometimes more optimistically called acceptance methods—retain the sampled values Y from g with a probability depending on Y. Thus

1. Generate Y from pdf g.
2. With probability $h(Y)$ return $X = Y$ else go to 1.

Then X has pdf proportional to gh. For

$$P(Y \leqslant x \text{ and } Y \text{ is accepted}) = \int_{-\infty}^{x} h(y)g(y)dy$$

so

$$P(Y \text{ is accepted}) = \int_{-\infty}^{\infty} h(y)g(y)dy$$

and

$$P(Y \leq x | Y \text{ is accepted}) = \int_{-\infty}^{x} g(y)h(y)dy \Big/ \int_{-\infty}^{\infty} gh \, dy$$

which shows the accepted values have pdf $gh/\int gh$.

We want to sample from f. Provided $f/g \leq M < \infty$ we can take $h = f/gM$, when X has pdf $f/M\int gh = f$. Furthermore, $P(Y \text{ is accepted}) = \int gh = 1/M$.

Algorithm 3.4 (general rejection). To sample from f with $f \leq Mg$.

    Repeat

        Generate Y from g,

        Generate U from $U(0, 1)$,

    until $MU \leq f(Y)/g(Y)$.

    Return $X = Y$.

The test $MU \leq f(Y)/g(Y)$ accepts Y with the required probability. The number of trials before a Y is accepted has a geometric distribution with mean M, so the algorithm works best if M is small. Note that it is not necessary to know f, only $f_1 \propto f$ and a bound on f_1/g.

Examples. (a) Beta distribution. Let $f_1(x) = x^{\alpha-1}(1-x)^{\beta-1}$ on $(0, 1)$, $Y \sim U(0, 1)$. Then f_1/g is bounded if and only if $\alpha, \beta \geq 1$. Then

$$M = (\alpha - 1)^{\alpha-1}(\beta - 1)^{\beta-1}/(\alpha + \beta - 2)^{\alpha+\beta-2}$$

which is near one only for both α and β small. If we take $g(x) = \alpha x^{\alpha-1}$, by $Y = U^{1/\alpha}$, then f_1/g is bounded for $\alpha > 0, \beta \geq 1$.

(b) von Mises distribution. Let $f_1(\theta) = (2\pi)^{-1} \exp(\kappa \cos \theta)$ on $(0, 2\pi)$, g uniform on $(0, 2\pi)$. Then $M = \exp \kappa \ (\kappa \geq 0)$ and the acceptance condition becomes $U \leq \exp[\kappa(\cos Y - 1)]$.

The art of using the rejection method is to find a suitable pdf g (known as the *envelope*) that matches f well and from which it is easy to sample. The theory applies equally well to discrete distributions, but suitable envelopes are very unusual in that case.

Distributions belonging to the exponential family often have $f_1(x) = \exp[-b(x)]$ for a particularly simple function $b(x)$. Von Neumann used a "trick" to generate an event with probability e^{-t}, $0 \leq t < 1$, which was subsequently exploited by Forsythe (1972). Consider the sequence

$$(t, U_1, U_2, \ldots)$$

for a sequence (U_i) of random numbers. Let N be the first index i with $U_i \geqslant U_{i-1}$. Then

$$P(N > n) = P(t > U_1 > \cdots > U_n) = t^n/n!$$

so

$$P(N = n) = P(N > n - 1) - P(N > n) = t^{n-1}/(n - 1)! - t^n/n!$$

and

$$P(N \text{ is odd}) = \sum_0^\infty P(N = 2k + 1) = \sum_0^\infty \{t^{2k}/2k! - t^{2k+1}/(2k + 1)!\} = e^{-t}$$

Thus we obtain

Algorithm 3.5 (Forsythe rejection). For pdf f with $b = -\ln(f/g)$, $0 \leqslant b(x) + \ln M \leqslant 1$ for all x.

1. Generate Y from g, let $U = b(Y) + \ln M$.
2. Generate U^*. If $U \leqslant U^*$ go to 4.
3. Generate U. If $U < U^*$ go to 2 else go to 1.
4. Return $X = Y$.

At each stage U is the last even term, and U^* is the last odd term in $(t, U_1, U_2, \ldots)$. The restriction $e^{-1} \leqslant f(x)/g(x)M \leqslant 1$ is severe and means that this method is never used alone.

Rejection may form a part of other algorithms. Consider Marsaglia's polar method, a modification of the Box–Muller Algorithm 3.1.

Algorithm 3.6 (polar). To generate two independent normal variates.

1. Repeat
   Generate $V_1, V_2 \sim U(-1, 1)$
   until $W = V_1^2 + V_2^2 < 1$.
2. Let $C = \sqrt{-2W^{-1} \ln W}$.
3. Return $X = CV_1$, $Y = CV_2$.

Step 1 is a rejection method leaving (V_1, V_2) uniformly distributed in the unit disc. Let (R, Θ) be polar coordinates for (V_1, V_2), so $W = R^2$. Then (W, Θ) has joint pdf $1/2\pi$ on $(0, 1) \times (0, 2\pi)$, whence W and Θ are uniform and independent. Let $E = -\ln W$. Then

$$X = \sqrt{2E} \cos \Theta = \sqrt{-2 \ln W}(V_1/\sqrt{W}) = CV_1$$

and similarly $Y = \sqrt{2E} \sin \Theta = CV_2$.

Algorithm 3.6 uses rejection to avoid calculating two trignometric functions and so is usually substantially faster than Algorithm 3.1, at the expense of a little extra complexity. It is folklore that the polar form is less susceptible to deficiencies in pseudo-random number generators. Figure 3.2 illustrates that this is not the case.

Kronmal and Peterson (1981, 1984) give methods related to rejection.

Composition

We can extend our techniques by first randomly choosing a distribution, then sampling from the chosen distribution. Suppose we have

$$f = \sum_{1}^{r} p_i f_i$$

for pdfs f_i and a probability distribution $\{p_1, \ldots, p_r\}$. Then we can sample from f by first choosing I from $\{p_i\}$, then taking a sample from f_I. The density f is said to be a *mixture* or *compound* of other distributions, and the method is known as *composition*.

One common use is to split the range of X up into intervals. For example, consider the standard exponential, and let f_i be the pdf conditional on $i - 1 \leq x < i$. Then $p_i = P(i - 1 \leq X < i) = e^{-(i-1)} - e^{-i} = e^{-(i-1)}(1 - e^{-1})$, so $\{p_i\}$ is a geometric distribution on $1, 2, 3, \ldots$. Furthermore,

$$f_i(x) = e^{-[x-(i-1)]}/(1 - e^{-1}) \text{ on } [i - 1, i)$$

Our outline algorithm is

1. $I = -1$
2. Repeat
   $I = I + 1$
   until (independent event with probability $1 - e^{-1}$).
3. Sample Y from pdf $e^{-x}/(1 - e^{-1})$ on $[0, 1)$.
4. Return $X = I + Y$.

Von Neumann noted that if we use a rejection method at step 3, we would accept with probability $(1 - e^{-1})$. Combining this observation with his "trick" gave

Algorithm 3.7 (von Neumann, exponential)

1. Let $I = 0$.
2. Generate U, set $T = U$.
3. Generate U^*. If $U \leq U^*$ return $X = I + T$.

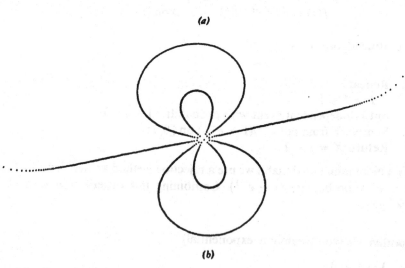

(a)

(b)

Figure 3.2. Plots of all pairs (X, Y) from the polar algorithm 3.6. (a) $X_i = (65X_{i-1} + 1)$ mod 2048. (b) $X_i = (1229X_{i-1} + 1)$ mod 2048.

4. Generate U. If $U < U^*$ go to 3.
5. $I = I + 1$, go to 2.

The reader is urged to check that this does indeed give samples from an exponential distribution. (See Exercise 3.7.)

Using rejection with the range of X split into parts was called "switching" by Atkinson and Whittaker (1976). A special case of their work is the beta distribution with $\alpha = \beta < 1$. We divide $(0, 1)$ into $(0, \frac{1}{2})$ and $(\frac{1}{2}, 1)$, using envelope $g_1(x) \propto x^{\alpha - 1}$ on $(0, \frac{1}{2})$ and $g_2(x) \propto (1 - x)^{\beta - 1}$ on $(\frac{1}{2}, 1)$. The envelope is sampled by

1. Let $Y = U_1^{1/\alpha}/2$.
2. Generate U_2. If $U_2 > \frac{1}{2}$ let $Y = 1 - Y$.

and the whole algorithm becomes

1. Repeat
    generate U_1, U_2, let $Y = U_1^{1/\alpha}/2$
    until $U_2 \leqslant [2(1 - Y)]^{\alpha - 1}$.
2. Generate U_3. If $U_3 > \frac{1}{2}$ let $Y = 1 - Y$.
3. Return $X = Y$.

using the symmetry of the density. Atkinson and Whittaker give a fuller version for $\alpha \neq \beta$.

Example. Brent (1974) applied composition and the von Neumann–Forsythe method to the normal distribution. The range $[0, \infty)$ is divided into intervals $I_i = [\Phi^{-1}(1 - 2^{-i}), \Phi^{-1}(1 - 2^{-i-1})]$, $i = 1, 2, 3, \ldots$. On each interval the trick can be applied since $\max f(x)/\min f(x) \leqslant e$ on each I_i. Interval I_i is selected with probability 2^{-i}, and finally a random sign is applied to the half-normal variate generated by composition. Brent gives a Fortran function which uses few uniforms per normal but has a large table of constants.

Example. Marsaglia and Bray (1964) used a four-part composition method for the normal distribution, which is discussed in more detail in Section 3.4. Two of the parts are very quick to sample and form 97.4% of the mixture. This illustrates the general point that composition algorithms tend to be fast but complex.

Ratio of Uniforms

Suppose (U, V) is a uniformly distributed point within the unit disc. From the polar algorithm V/U has the distribution of the ratio of two independent normal variates. From the Box–Muller algorithm this is the distribution of $\tan \Theta$, which is easily seen to be the Cauchy distribution. Thus a simple way to sample from the Cauchy is

Algorithm 3.8 (Cauchy)

> **Repeat**
> > generate $U_1, U_2 \sim U(0, 1)$,
> > let $V = 2U_2 - 1$
> **until** $U_1^2 + V^2 < 1$.
> **Return** $X = V/U_1$.

Here (U_1, V) is uniform within a semicircle, which clearly suffices.

Kinderman and Monahan (1977) took up this idea and considered whether other distributions could be sampled as V/U for (U, V) uniform over some set.

Theorem 3.2. For any nonnegative function h with $\int h < \infty$ let $C_h = \{(u, v) \mid 0 \leqslant u \leqslant \sqrt{h(v/u)}\}$. Then C_h has a finite area and if (U, V) is uniformly distributed over C_h then $X = V/U$ has pdf $h/\int h$.

PROOF. Consider the change of variables $(u, v) \to (u, x = v/u)$. Now

$$\text{area } (C_h) = \iint_{C_h} du\,dv = \iint_0^{\sqrt{h(x)}} u\,du\,dx = \int \tfrac{1}{2} h(x) dx < \infty$$

Furthermore, (U, V) has pdf $1/\text{area}(C_h)$, so (U, X) has pdf $u/\text{area}(C_h)$ and X has marginal pdf

$$\int_0^{\sqrt{h(x)}} u\,du/\text{area}(C_h) = h(x)/2\,\text{area}(C_h) = h(x)/\int h(x)$$

as required.   □

This result is most useful when C_h is contained in a rectangle $[0, a] \times [b_-, b_+]$, when we can use rejection sampling.

Algorithm 3.9 (ratio of uniforms)

> **Repeat**
> > generate $U_1, U_2 \sim U(0, 1)$,
> > let $U = aU_1, V = b_- + (b_+ - b_-)U_2$.
> **until** $(U, V) \in C_h$.
> **Return** $X = V/U$.

Note that as in rejection sampling we only need to know f up to a constant factor. Conditions for C_h to be contained in a rectangle are given by

Theorem 3.3. Suppose $h(x)$ and $x^2 h(x)$ are bounded. Then $C_h \subset [0, a] \times [b_-, b_+]$ where $a = \sqrt{\sup h}$, $b_+ = \sqrt{\sup\{x^2 h(x) | x \geqslant 0\}}$ and $b_- = -\sqrt{\sup\{x^2 h(x) | x \leqslant 0\}}$.

PROOF. $0 \leqslant u \leqslant \sqrt{h(v/u)} \leqslant \sqrt{\sup h}$ is obvious. For $v \geqslant 0$ to be a possible value there must exist a $u > 0$ with $0 < u^2 \leqslant h(v/u)$ or $t > 0$ with $v^2 \leqslant t^2 h(t)$ for $t = v/u$. Thus $(u, v) \in C_h$ implies $v^2 \leqslant b_+^2$ or $v \leqslant b_+$. The case $v < 0$ follows similarly. $\qquad\square$

It may be possible to enclose C_h in other polygonal shapes more efficiently than within a rectangle. However, the main computation is usually in checking if $(u, v) \in C_h$.

Examples. (a) Exponential. Let $h(x) = e^{-x}$ on $(0, \infty)$. Then $a = 1$, $b_- = 0$, and $b_+ = 2/e$. Furthermore, $(u, v) \in C_h$ is equivalent to $u^2 \leqslant e^{-v/u}$ or $v \leqslant -2u \ln u$. Thus the algorithm is

    Repeat
        generate U, $V = 2U_1/e$
    until $V \leqslant -2U \ln U$.
    Return $X = V/U$.

(b) Cauchy. Take $h(x) = 1/(1 + x^2)$ on $(-\infty, \infty)$. Then $a = 1$ and $b_+ = b_- = 1$. Also $(u, v) \in C_h$ is $u^2 \leqslant 1/(1 + v^2/u^2)$ or $v^2 + u^2 \leqslant 1$, so we recover Algorithm 3.8.

(c) Normal. Let $h(x) = \exp(-\frac{1}{2}x^2)$. Then $a = 1$, $b_+^2 = b_-^2 = 2/e$, and $(u, v) \in C_h$ if and only if $v^2 \leqslant -4u^2 \ln u$. The algorithm is

    Repeat
        generate U, U_1,
        let $V = \sqrt{2e^{-1}}(2U_1 - 1)$, $X = V/U$
    until $X^2 \leqslant -4 \ln u$.

Figure 3.3 shows how this transforms the lattice structure of pseudo-random numbers.

Squeezing

Both the rejection and ratio-of-uniforms methods use membership tests like $MU \leqslant f/g$ or $(U, V) \in C_h$ which can be slow to evaluate. Most of the time the test will be clearly passed or failed and simple approximations to f/g or C_h will suffice.

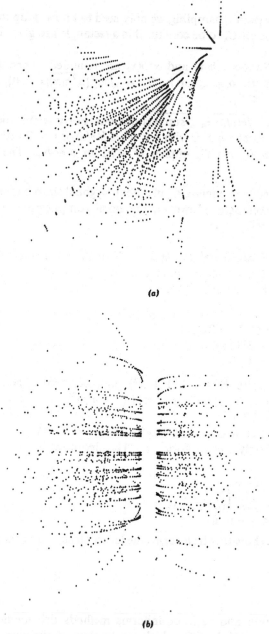

(a)

(b)

Figure 3.3. Plots of all possible successive normals (X, Y) from Algorithm 3.17. *(a)* $X_i = (65X_{i-1} + 1) \bmod 2048.$ *(b)* $X_i = (1229X_{i-1} + 1) \bmod 2048.$

Suppose we can find simple (to evaluate) functions l and u with

$$l \leqslant f/g \leqslant u \quad \text{for all } x$$

Then if $MU \leqslant l(Y)$ we can accept Y, and if $MU > u(Y)$ we can reject Y, in each case *without* evaluating f/g. This can result in a considerable speeding-up of the algorithm. The analogue for the ratio of uniforms is to find sets $C_l \subset C_h \subset C_o$ with $(u, v) \in C_l$ and $(u, v) \notin C_o$ being easy to determine.

This process is called *pretesting* or *squeezing* (Marsaglia, 1977). Much ingenuity can be applied to the choice of the bounds and even the order of the tests.

Examples. (a) Exponential distribution restricted to $(0, 2)$, with pdf $f(x) = e^{-x}/(1 - e^{-2})$. With g the uniform density on $(0, 2)$ we obtain

> Repeat
>   generate $Y \sim U(0, 2)$, $U \sim U(0, 1)$
>   until $U \leqslant e^{-Y}$.
> Return $X = Y$.

from $f_1(x) = e^{-x}$. We can then apply squeezing to $U \leqslant e^{-Y}$. Of course, $e^x \geqslant 1 + x$ for all x, so $1 - x \leqslant e^{-x} \leqslant 1/(1 + x)$, and so

$$e^{-a}(a + 1 - x) \leqslant e^{-x} \leqslant e^{-a}/(1 - a + x)$$

The squeezed rejection algorithm is

1. Generate $Y \sim U(0, 2)$, $U \sim U(0, 1)$.
2. If $U \leqslant e^{-a}(a + 1 - Y)$ go to 5.
3. If $U > e^{-b}/(1 - b + Y)$ go to 1.
4. If $U > e^{-Y}$ go to 1.
5. Return $X = Y$.

where a and b are to be chosen. We choose a to maximize the probability p that the test succeeds. We find $p = ae^{-a}$ for $a \geqslant 1, \frac{1}{4}(a + 1)^2 e^{-a}$ for $a < 1$ and so take $a = 1$. We choose b to maximize the probability that we branch to 1; this is $\frac{1}{2}[3 - b - e^{-b}(1 + \ln(3 - b) + b)] \approx 0.521$ at $b = 0.662$. With these choices 36.8% of the time Y is accepted at 2, 52.1% we reject at 3, 8.9% we reject at 4, and 6.4% we accept at 4. Figure 3.4 illustrates the pretests.

(b) Ratio-of-uniforms for exponential. We want to pretest $V \leqslant -2U \ln U$. From $e^x \geqslant 1 + x$ we obtain $x \leqslant \ln(1 + x)$ or $y - 1 \geqslant \ln y$ and $y^{-1} - 1 \leqslant -\ln y$, so

$$(1 + \ln a) - aU \leqslant -\ln U \leqslant a/U - (1 + \ln a)$$

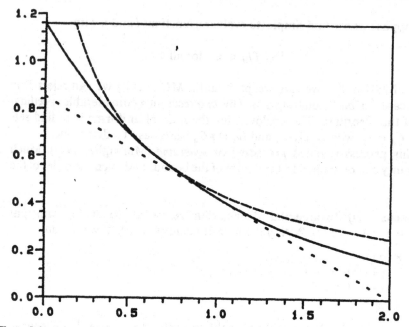

Figure 3.4. Pretests for the exponential distribution on (0, 2). The solid line is the pdf; the dashed lines are the envelopes.

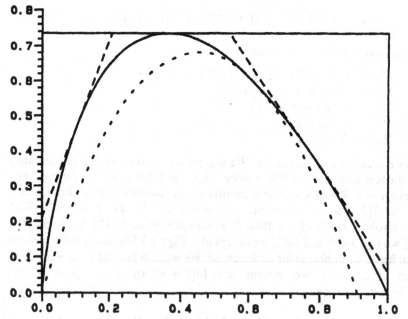

Figure 3.5. Pretests for the ratio method for the exponential. The solid line is C_h; the dashed lines are the pretests.

We choose constants a, b_1, b_2 in

1. Generate $U \sim U(0, 1)$, $V \sim U(0, 2/e)$.
2. Let $X = V/U$.
3. If $\frac{1}{2}X \leqslant (1 + \ln a) - aU$ go to 6.
4. If $\frac{1}{2}X > b_1/U - (1 + \ln b_1)$ go to 1.
5. If $\frac{1}{2}X > b_2/U - (1 + \ln b_2)$ go to 1.
6. Return X.

The probability of acceptance at 3 is $e(1 + \ln a)^3/6a^2$, maximized for $a \approx 1.65$. The presence of both 4 and 5 corresponds to two possible choices for b of 0.105 and 0.773. Figure 3.5 shows the bounds.

3.3. DISCRETE DISTRIBUTIONS

We will assume throughout this section that we wish to sample a variate X from a distribution given by $p_r = P(X = r)$, $r = 1, 2, \ldots$ and $P_r = P(X \leqslant r)$. Any distribution on a countable set can be reduced to this form by relabeling the points. We will assume that the number of points is M, possibly infinity. Of course, for any method that involves storing (p_r) we will have to truncate a distribution with M extremely large or infinite, say by ignoring all values beyond M' where $1 - P(X \leqslant M') \leqslant \varepsilon$, say 10^{-6}.

Inversion

For a discrete distribution $F^-(u) = \min\{x|F(x) \geqslant u\} = i$ where $P_{i-1} < u \leqslant P_i$, so inversion amounts to searching a table of (P_i) for a suitable index i. Formally we have:

Algorithm 3.10

1. Generate $U \sim U(0, 1)$. Let $i = 1$.
2. While $P_i \leqslant U$ do $i = i + 1$.
3. Return $X = i$.

The expected number of comparisons at step 2 is EX, since i comparisons are done for $X = i$. The algorithm can be speeded up by reordering the (p_r) into decreasing order. This reduces EX as much as possible, and the original distribution can be recovered. The expense is in set-up time and space.

A better way to reduce the number of comparisons is to start the search at a more suitable place. If (p_r) is unimodal one could search left or right from the mode. We can also use a binary search to locate i.

Algorithm 3.10A

1. Generate $U \sim U(0, 1)$, set $L = 0, R = M$.
2. Repeat
    $$i = \text{int}[(L + R)/2]$$
    if $U > P_i$ then $L = i$ else $R = i$
    until $L \geqslant R - 1$.
3. Return $X = i$.

This tends to be faster for $M \geqslant 30$.

More generally we can start the search from a point depending on U. The indexed search method is based on the use of a thumb-index in a dictionary to find the beginning of the L's, say (Chen and Asau, 1974).

Algorithm 3.11 (indexed search). Fix m. Let $q_j = \min\{i \mid P_i \geqslant j/m\}$, $j = 0, \ldots, m - 1$.

1. Generate U. Let $k = \text{int}(mU), i = q_k$.
2. While $P_i \leqslant U$ do $i = i + 1$.
3. $X = i$.

This is verified as for Algorithm 3.10 plus noting that initially $P_{i-1} < j/m \leqslant P_i$, so if $P_i \geqslant U, P_i \geqslant U \geqslant j/m > P_{i-1}$.

Yet more sophisticated search algorithms are possible, the process being a trade-off between the time to set up additional structures and the time taken for each call.

Example. $[X \sim \text{Poisson (10)}]$. We find $P(X \leqslant 22) \approx 0.9997$ and may truncate the table there, so possible values are $0, \ldots, 22$. For searching from 0 we expect 11 comparisons $[= E(X + 1)$ to conform to a range $1, \ldots, M]$. Searching from the mode reduces this to about 3.6, and an indexed search with $m = 5$ to about 3.3 plus 1 look-up in the q table. Figure 3.6 shows a binary search which needs about 3.3 comparisons per X. The tree is found by combining the two least probable groups of values at each stage, starting with the individual values (Knuth, 1973a, p. 402).

Ahrens and Kohrt (1981) modify the indexed search idea by subdividing $\{q_{m-1}, \ldots, 1\}$ by a further index. This may be faster for long-tailed distributions. They also record if $i = q_j$ is the *only* value with $j/m \leqslant P_i < (j + 1)/m$.

Alias Method

Walker (1974, 1977) proposed what at first sight is an ingenious modification of the rejection method for discrete distributions, but is in fact a composition method. Instead of rejecting $X = j$ we output $X = A(j)$, the *alias* value.

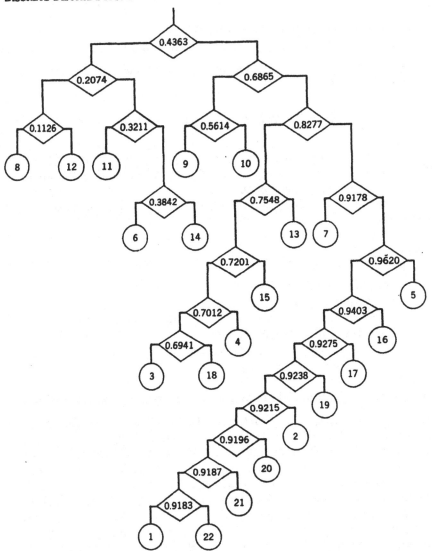

Figure 3.6. Binary search tree for Poisson (10). At each node move left if $U \leqslant$ stated value, right otherwise.

The method requires that M be finite, and uses two tables Q and A of probabilities and aliases respectively.

Algorithm 3.12 (alias method)

   1. Generate $U \sim U(0, 1)$.

2. Let $Y = 1 + \text{int}[MU]$, $Z = \text{frac}[MU]$.

3. If $Z \leq Q(Y)$, return $X = Y$ else return $X = A(Y)$.

At step 2 Y is uniformly distributed on $\{1, \ldots, M\}$ and $Z \sim U(0, 1)$, independent of Y. Thus

$$P(X = i) = Q(i)/M + \sum_{j:A(j)=i} [1 - Q(j)]/M \qquad (1)$$

The computation of (3.12) can be speeded up by replacing $Z \leq Q(Y)$ by $MU \leq Q(i) + i - 1$ or $U \leq [Q(i) + i - 1]/M$ and tabulating the appropriate right-hand side in place of Q.

It remains to find tables Q and A so that $P(X = r) = p_r$. These tables are not unique.

Algorithm 3.13

> For $i = 1$ to M
> > Set $Q(i) = 1$, $a_i = p_i$, $I_i = $ true.
> > For step $= 1$ to $M - 1$ do
> > > Select i with $a_i < 1/M$, $I_i = $ true. If there is none, stop.
> > > Select j with $a_j > 1/M$, $I_j = $ true.
> > > Set $I_i = $ false, $A(i) = j$, $Q(i) = Ma_i$.
> > > Set $a_j = a_j - [1 - Q(i)]/M$.

Theorem 3.4. Algorithm 3.13 finds Q and A satisfying (1) for $p_r = P(X = r)$.

PROOF. At all times $a_i = p_i - $ last term of (1). When I_i is set false, $Q(i)/M = a_i$ by definition, so $P(X = i) = p_i$. Equation (1) for i is unchanged subsequently. Let $d_i = a_i - Q(i)/M$. Then initially $\sum d_i = 0$, and this is unchanged at each step. For indices i with $I_i = $ false we have $d_i = 0$, so $\sum' d_i = 0$, the sum being over indices with $I_i = $ true. At each stage either there are indices i and j with $d_i < 0$, $d_j > 0$ or all $d_i = 0$ and we are done. After $M - 1$ steps, only one $I_i = $ true, so that $d_i = 0$. It remains to check that $0 \leq Q(i) \leq 1$. The upper inequality follows from the choice of i. Initially $a_i \geq 0$, and by the choice of j, $a_j = a_j - [1 - Q(i)]/M \geq 1/M - 1/M = 0$ at the end of each step. $\square$

This proof shows that an explicit choice of i and j can be given by

Algorithm 3.13A

> For $i = 1$ to M
> > Set $Q(i) = 1$, $a_i = p_i$, $I_i = $ true.
> > For step $= 1$ to $M - 1$
> > > Choose i attaining $\min\{a_i | I_i \text{ true}\}$.
> > > Choose j attaining $\max\{a_j | I_j \text{ true}\}$.

If $a_i = a_j$ stop.
Set I_i = false, $Q(i) = Ma_i$, $A(i) = j$.
Set $a_j = a_j - [1 - Q(i)]/M$.

Example. [$X \sim$ binomial (3, 1/3)]

	0	1	2	3		0	1	2	3
$108a$	32	48	24	4		32	25	24	4
A	—	—	—	—		—	—	—	—
$27Q$	27	27	27	27	→	27	27	27	4
I	T	T	T	T		T	T	T	F

		0	1	2	3		0	1	2	3
	$108a$	29	25	24	4		27	25	24	4
	A	—	—	0	1		—	0	0	1
→	$27Q$	27	27	24	4	→	27	25	24	4
	I	T	T	F	F		T	F	F	F

Unfortunately this version of Algorithm 3.13 uses $O(M + 1\text{-step})$ operations to find i and j, and so $O(M^2)$ operations overall. An $O(M)$ implementation is possible, using linked lists (Kronmal and Peterson, 1979) or pointers (Greenwood, 1981). We illustrate the use of stacks.

Algorithm 3.13B. Needs work array $w(1) \ldots w(M)$ of indices

Set $nn = 0$, $np = M + 1$.
For $i = 1$ to M
  Set $Q(i) = Mp_i$.
  If $Q(i) < 1$ set $nn = nn + 1$, $w(nn) = i$
    else set $np = np - 1$, $w(np) = i$.
For step $= 1$ to $M - 1$
  $i = w(\text{step})$, $j = w(np)$
  $A(i) = j$, $Q(j) = Q(j) + Q(i) - 1$
  If $Q(j) < 1$ then $np = np + 1$.

In this implementation $Q(i)$ stores Ma_i and indices with $a_i < 1/M$ are step $\ldots np - 1$, the rest having $a_i \geqslant 1/M$.

Table 3.1 shows that Algorithm 3.13B can be much faster than Algorithm 3.13A. However, Algorithm 3.13A tries to choose i and j to maximize the $Q(i)$ and hence minimize look-ups of aliases. For example, if in the binomial

Table 3.1.   Timings on BBC Microcomputer of Methods for the Poisson
Distribution (Interpreted BASIC)

	μ			
	5	10	20	50
Algorithm 3.3				
Set up (msec)	14	22	37	85
Per call (msec)	33	59	109	260
Straight search				
Set up (msec)	180	280	470	910
Per call (msec)	15	26	49	114
Per call—3.10A (msec)	26	30	34	40
Indexed search ($m = 2\mu$)				
Set up (msec)	280	470	820	1,780
Per call (msec)	8.2	8.1	8.1	7.7
Mean comparisons	1.56	1.51	1.42	1.61
Binary search				
Set up (sec)	1.4	2.7	6.2	22.4
Per call (msec)	15	18	20	22
Alias				
Per call (msec)	6.8	6.8	6.8	6.8
Set up—3.13A (sec)	2.08	4.50	10.4	38.0
Set up—3.13B (sec)	0.58	0.89	1.47	2.79
Algorithm 3.15				
Set up (msec)	—	56	57	57
Per call (msec)	—	132	125	169

example we always take i, j as small as possible we find

	0	1	2	3
A	1	—	0	0
$27Q$	6	27	24	4

so the probability of choosing an alias is increased from 7/27 to 23/54 and
the time per call will be increased slightly.

Table Method

This is an approximate method suggested by Marsaglia (1963) and
implemented by Norman and Cannon (1972). Consider the d-digit radix r

representation of p_i,

$$p_i = \sum_1^d r^{-j} a_{ij}$$

For example, we have the following decimal expansions for binomial $(3, 1/3)$:

$$P(X = 0) = 0.296$$
$$P(X = 1) = 0.445 \text{ (rounded up to make total 1)}$$
$$P(X = 2) = 0.222$$
$$P(X = 3) = 0.037$$

The method depends on generating events of probability r^{-j} and assigning a_{ij} of them to $X = i$. In our example we have

1. Generate $U \sim U(0, 1)$.
2. If $0 \leqslant U < 0.8$, let $I = \text{int}[10U] + 1$, $X = a_1[I]$.
   If $0.8 \leqslant U < 0.98$, let $I = \text{int}[100U] - 80 + 1$, $X = a_2[I]$.
   If $0.98 \leqslant U$, let $I = \text{int}[1000U] - 980 + 1$, $X = a_3[I]$.

Here a_j is a table of a_{0j} 0's, a_{1j} 1's, and so on for $j = 1, 2$, or 3. If $0 \leqslant U < 0.8$, I is uniform on $1, \ldots, 2 + 4 + 2 + 0$, and if $0.8 \leqslant U < 0.98$, I is uniform on $1, \ldots, 9 + 4 + 2 + 3$, and if $0.98 \leqslant U$, I is uniform on $1, \ldots, 6 + 5 + 2 + 7$. Thus $P(X = 0) = 0.8 \times 2/8 + (0.98 - 0.8) \times 9/18 + (1 - 0.98) \times 6/20 = 0.2 + 0.09 + 0.006$. The principle should be clear from this example. Peterson and Kronmal (1983) give a formal algorithm. Comparative studies have shown the table method to have slow set-up times but fast sampling. It seems rarely to be competitive with the search and alias methods.

Specific Distributions

The general methods of indexed search and alias are easy to implement for an arbitrary discrete distribution, so methods specific to particular distributions seem only worthwhile if the set-up times are significant, for example if the parameters change every few samples.

Geometric Distribution

The geometric distribution is sometimes known as the "discrete exponential" and can be sampled by discretizing an exponential. Suppose E has an exponential (λ) distribution, and $X = \text{int}[E]$. Then

$$P(X = r) = P(r \leqslant E < r + 1) = e^{-\lambda r} - e^{-\lambda(r+1)} = (e^{-\lambda})^r (1 - e^{-\lambda})$$

for $r = 0, 1, 2, \ldots$. This is a geometric distribution with success probability $p = 1 - e^{-\lambda}$, so choosing $\lambda = -\ln(1 - p)$ we can sample from any given geometric distribution.

A negative binomial distribution with integer index k is the distribution of the sum of k independent geometric variates. It can be sampled as such for small k or k varying from sample to sample.

Binomial Distribution

A binomial(n, p) distribution is the distribution of the sum of n Bernoulli trials. If we sample the trials by inversion we generate $U_1, \ldots, U_n$ and count the number Y of U_i less than p.

For large n it will be wasteful to generate all the U_i to find $U_{(Y)} \leqslant p < U_{(Y+1)}$, where $U_{(i)}$ is the ith smallest U_i. We can make use of the fact that $U_{(i)}$ has a beta(i, $n + 1 - i$) distribution, and that conditional on $U_{(i)} \leqslant p < U_{(i+k)}$, $Y - i$ has a binomial($k - 1$, $[p - U_{(i)}]/[U_{(i+k)} - U_{(i)}]$) distribution. Relles (1972) applied this recursively, starting with $(0, 1)$ and in each case generating the median of the range enclosing p. (He gives Fortran code using an approximate method to generate beta variates.) Knuth (1981, p. 131) gives a more refined version due to Ahrens which for binomial(n, p) chooses $i = \text{int}[(1 + n)p]$ then works recursively until k is small. A formal algorithm is

Algorithm 3.14

> Set $k = n, \theta = p, X = 0$.
> Repeat
> $i = \text{int}[1 + k\theta]$
>   Generate $V \sim \text{beta}(i, k + 1 - i)$
>   If $\theta < V$ then $\theta = \theta/V, k = i - 1$
>   else $X = X + i, \theta = (\theta - V)/(1 - V), k = k - i$
> until $k \leqslant K$.
> For $i = 1$ to k
>   generate $U \sim U(0, 1)$; if $U < p$ then $X = X + 1$.
> Return X.

Here K is chosen as a balance for efficiency (Exercise 3.13).

Another stratagem is available if p is a simple binary fraction (Tocher, 1954). If $p = \frac{1}{2}$, we can generate n random *bits* and count the number which are one. If $U \sim U(0, 1)$ its binary representation is such a series of random bits. If $p = 2^{-r}$, we can take $U_1, \ldots, U_r$ and count the number of bits that are one in all U_i, hence in U_1 OR U_2 OR $\cdots$ OR U_r. It is easy to extend the method to events of probability $m2^{-r}$. For example, if $p = \frac{5}{8}$ we count the bits that are either one in U_1 or one in both U_2 and U_3, events of probability $\frac{5}{8}$, so Y is the number of one's in U_1 OR (U_2 AND U_3). This method is not recommended because the bits of pseudo-random numbers are not usually a good source of random bits.

Poisson Distribution

Table 3.1 shows that the simple Algorithm 3.3 with expected time proportional to $\mu + 1$ is slow unless set-up time is significant. Atkinson (1979a, 1979b) and Peterson and Kronmal (1983) give more extensive comparisons which favor the general indexed search and alias methods.

The general methods need large tables for μ large (say 100), and there has been interest in different methods for large μ. Atkinson (1979a), Ahrens and Dieter (1980, 1982a), and Devroye (1981) all give methods that use rejection in the tail. All are fairly complex methods: Devroye gives a 65-line Fortran routine. The obvious idea to use a normal distribution as the envelope does not work as the normal tail decays too fast. Atkinson used a logistic envelope. His algorithm is

Algorithm 3.15 [Poisson (μ), $\mu \geqslant 30$]

> Set $c = 0.767 - 3.36/\mu$, $\beta = \pi(3\mu)^{-1/2}$, $\alpha = \beta\mu$,
> $\quad k = \ln c - \mu - \ln \beta$.
> Repeat
> $\quad$ Repeat
> $\qquad$ generate U_1,
> $\qquad$ Let $X = (\alpha - \ln[(1 - U_1)/U_1])/\beta$
> $\quad$ until $X > -\frac{1}{2}$
> $\quad$ Let $N = \text{int}[X + \frac{1}{2}]$.
> $\quad$ Generate U_2.
> until $\alpha - \beta X + \ln\{U_2/[1 + \exp(\alpha - \beta X)]^2\} \leqslant k + N \ln \mu - \ln N!$
> Return N.

The restriction $\mu \geqslant 30$ ensures (it is claimed) that the envelope constant c suffices. The method is faster than Algorithm 3.3 from about $\mu = 30$ and comes into its own for $\mu \geqslant 100$.

A further possibility (Atkinson, 1979a) for a general Poisson algorithm is to have available fast generators for particular values of μ, say $4, 8, 16, \ldots, 128$ and to use the fact that the sum of independent Poisson variates is Poisson. Thus $Y \sim \text{Poisson}(84.7)$ would be generated as $Y_1 + Y_2 + Y_3 + Y_4$ where $Y_1 \sim \text{Poisson}(64)$, $Y_2 \sim \text{Poisson}(16)$, $Y_3 \sim \text{Poisson}(4)$, and $Y_4 \sim \text{Poisson}(0.7)$ would be generated by Algorithm 3.3. Such an algorithm would be bulky and complex but desirable when μ changed every call, for which we will see uses in Chapter 4.

Sampling without Replacement

Some discrete distribution problems will require sampling without replacement, particularly for randomization of experiments and surveys. Suppose we wish to draw n units at random from a population of N, for example n individuals from a database. The obvious method is to select individuals at random, rejecting those already selected until n are found. If n is comparable with N, this is wasteful. One way to avoid this is to generate $U_1, \ldots, U_N \sim U(0, 1)$ and select the n individuals corresponding to the n smallest U_i, perhaps found by sorting (U_i, i) on U_i (Page, 1967). In either case the selected individuals could then be picked out of the database sequentially.

A better idea was given independently by Fan et al. (1962), Jones (1962), and Bebbington (1975). The individuals are considered in turn. If k have already been seen and r chosen, the next is selected with probability $(n - r)/(N - k)$. This terminates when n are chosen, often before the Nth. It is not immediately obvious that this method will pick n individuals, nor that it picks them at random. However, the sampling probability reaches one when there are $n - r$ individuals left, so n will be picked. Furthermore, the probability that elements $i_1 < i_2 < \cdots < i_n$ are picked is

$$\prod_1^N R_t/[N - t + 1]$$

where $R_{i_s} = n - s + 1$ and $R_t = (N - t + 1) - (n - s)$ for $i_s < t < i_{s+1}$. The denominator is $N!$ and the numerator contains $n, n - 1, \ldots, 1$ for $i_1, \ldots, i_s$ and $N - n, \ldots, 1$ for the remaining elements. Thus the probability is $n!(N - n)!/N!$ independent of which elements are specified. See also Vitter (1984) who samples $i_{j+1} - i_j$ directly.

Knuth (1981) and McLeod and Bellhouse (1983) give the following method, which McLeod and Bellhouse show to be marginally faster than the previous method and which does not require N to be known before the database is read. Select the first n individuals as a current sample. Each subsequent individual is rejected with probability $1 - n/t$, where t is the number of individuals seen. If selected, the new individual replaces one of the current sample chosen at random. Clearly this will select n individuals. For $t \geq n$, the tth individual is selected with probability n/t and survives with probability

$$\prod_{t+1}^N \left(1 - \frac{1}{r}\right) = \frac{t}{N}$$

since $1/r$ is the probability that the rth individual will replace it. By reordering the first n individuals, we see *any* individual is selected with probability n/N. Now for individuals $t_1 < t_2 < \cdots < t_n$ to be selected we have probability

$$\prod_1^n \left[\frac{n - i + 1}{t_i} \prod_{t_i+1}^{t_{i+1}-1} \left(1 - \frac{i}{r} \right) \right] = n! \prod_1^n \frac{(t_i - 1)!/(t_i - i)!}{s_i!/(s_i - i)!} = \frac{n!(N - n)!}{N!}$$

where $t_{n+1} = N + 1$ and $s_i = t_{i+1} - 1$, by checking that the t_ith element is selected and does not displace $t_1, \ldots, t_{i-1}$, and that elements with $t_i < t < t_{i+1}$ do not replace any of the i so far selected. This formula is derived for $t_1 \geqslant n$ but is easily modified to cover the remaining cases (Exercise 3.14).

Occasionally there will be a need for a random permutation, equivalent to shuffling N objects. In principle this is easy: label all $N!$ permutations and choose one at random. The following algorithm is due to Moses and Oakford (1963).

Algorithm 3.16

> For $t = N$ down to 1 do
> > Generate S uniform on $\{1, \ldots, t\}$.
> > Swap the Sth and tth objects.

To see that this generates all permutations, define

$$p = \prod_{t=1}^N (t - 1)!(S_t - 1)$$

where S_t is the value of S on the tth step. Then $0 \leqslant p < N!$ and all values occur. Further p determines the (S_t) and hence the permutation. Finally, all values of p are equally likely. Algorithm 3.16 avoids having to generate a digit in $1, \ldots, N!$, which can be a problem for even moderately large N. It has been rediscovered frequently (Durstenfeld, 1964; de Balbine, 1967; Page, 1967).

3.4. CONTINUOUS DISTRIBUTIONS

For discrete distributions we saw that general algorithms were competitive for most specific distributions. This is not true for continuous distributions, perhaps because of their many different tail behaviors. A very large number of specialized algorithms have been proposed, and we consider in detail only the more popular or meritorious ones.

Normal Distribution

The normal distribution is a location-scale family, so by letting $X = \mu + \sigma Y$ we can concentrate on $Y \sim N(0, 1)$. We have already seen four methods: sum of 12 uniforms, Box–Muller (Algorithm 3.1), Marsaglia's polar variant (Algorithm 3.6), and Brent's von Neumann–Forsythe method.

A more recent method is the ratio-of-uniforms (Best, 1979; Knuth, 1981, pp. 125–127, 552; Ripley, 1983c). Take $h = \exp(-x^2/2)$. Then $(u, v) \in C_h$ is equivalent to $u \geqslant 0$, $u^2 \leqslant \exp(-v^2/2u^2)$ or $-4u^2 \ln u \geqslant v^2$ or $x^2 \leqslant -4 \ln u$ for $x = v/u$. With pretests this gives

Algorithm 3.17 (normal by ratio-of-uniforms)

1. Generate U, $V = \sqrt{(2/e)}(2U_1 - 1)$. $[V = 0.86(2U_1 - 1)]$
2. Let $X = V/U$, $Z = \frac{1}{4}X^2$.
3. If $Z < (1 + \ln a) - aU$ go to 6. $[Z < 1 - U$ recommended$]$
4. If $Z > b/U - (1 + \ln b)$ go to 1. $[Z > 0.259/U + 0.35]$
5. If $Z > -\ln U$ go to 1.
6. Return X.

Here $b_+^2 = b_-^2 = \max(x^2 e^{-x^2/2}) = 2/e$ so $b_+ = \sqrt{2/e} \leqslant 0.86$. Best chooses $a = 1$, and this seems faster than Knuth's $a = e^{1/4}$ which minimizes the probability of reaching 5. For b, $e^{-1.35} \approx 0.259$ is near-optimal for avoiding 5. This algorithm is as simple as the polar method, returns one normal deviate at a time, and is comparable in speed. (See Table 3.2.) Figure 3.7 illustrates the acceptance region and pretests. Figure 3.3 shows all possible successive pairs (X, Y), to be compared with Fig. 3.1 and 3.2. There is little to choose between the Box–Muller, polar, and ratio methods for sensitivity to pseudo-random number generators.

Compilers of subroutine libraries will be prepared to accept more complex algorithms for the sake of speed. Brent's algorithm is just one of a number of composition algorithms designed for speed. Marsaglia and Bray (1964) gave a four-part algorithm. Density f_4 copes with $|X| \geqslant 3$ whereas f_1, f_2, f_3 all have support $(-3, 3)$. Thus $p_4 = P(|X| \geqslant 3) \approx 0.0027$. In the polar algorithm $|X| \geqslant 3$ can only occur if $-2 \ln W \geqslant 9$. This yields the following algorithm for f_4:

Repeat
  Repeat
      Generate $V_1, V_2 \sim U(-1, 1)$
      until $W = V_1^2 + V_2^2 < 1$.
    Let $C = \sqrt{W^{-1}(9 - 2 \ln W)}$, $S = CV_1$, $T = CV_2$.
  until $|S| > 3$ or $|T| > 3$.
  If $|S| > 3$ return $X = S$ else return $X = T$.

Table 3.2. Comparisons of Various Methods for Sampling Nonuniform Distributions (Times in msecs for Fortran Except Where Stated)

	BBC Basic	ACT Sirius		Corvus	VAX(μsec)	CDC(μsec)		APL on IBM 370/168[c]
						RANF	G05CAF	
	1.5	4.0[a]	1.0[b]	91μs	29	2.8	30	29
Uniform								
Exponential								
Inversion	18	11	5.0	3.1	95	80	110	89
Von Neumann	14	30	3.5	0.92	155	50	180	—
ratio	22	25	4.8	6.9	155	110	160	—
Cauchy								
Inversion	24.5	38	8.0	3.9	125	120	145	107
ratio	16.5	28	4.8	1.6	110	45	130	178
Normal								
Box–Muller	42	32	7.0	7.7	145	160	190	127
polar	31	27	6.0	4.1	115	82	120	134
ratio	32	44	7.8	3.2	145	80	155	223
Brent	44	66	8.9	3.0	115	90	130	—
Marsaglia–Bray	17	—	—	1.45	155	55	160	—
inversion	36	—	—	4.0	125	110	140	323
Gamma								
α < 1 (3.19)	50–95	ca. 50	ca. 18	13–16	320–390	230–270	310–390	ca. 350
α > 1 (3.20)	50–70	ca. 80	ca. 20	8–10	230–270	180–210	270–300	300–350

[a]Interpreted Basic.
[b]Compiled Basic.
[c]For 1000 samples. Timings by Dr. P. J. Green.

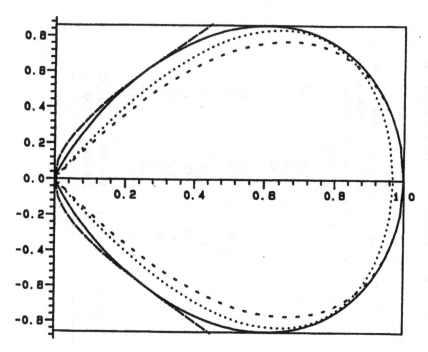

Figure 3.7. Acceptance region and pretests for Algorithm 3.17. The region C_k is inside the solid line, the dashed lines are the recommended pretests, and the dotted line is $Z = 1.25 - e^{1/4}U$.

On $(-3, 3)$ the normal pdf is approximated by sums of two or three uniforms. Thus

$$p_1 = 0.8638, \quad X = 2(U_1 + U_2 + U_3) - 3$$
$$p_2 = 0.1107, \quad X = 1.5(U_1 + U_2 - 1)$$
$$p_3 \approx 0.0228, \quad f_3 = [\phi - p_1 f_1 - p_2 f_2]/P(|X| \leq 3)$$

The reader may wonder how these values came to be chosen. Presumably $p_1 f_1$ was chosen to be a good, simple approximation to ϕ, p_1 being chosen so that $p_1 f_1 \leq \phi$ for all x. A plot of $\phi - p_1 f_1$ shows a roughly triangular middle part, which suggests the form of f_2. Then f_3 and f_4 were chosen to fill in the remaining pieces. As they are sampled infrequently, this can be done without undue regard for efficiency. See Figure 3.8.

The complete algorithm is

Algorithm 3.18 (Marsaglia–Bray, normal)

Generate U.

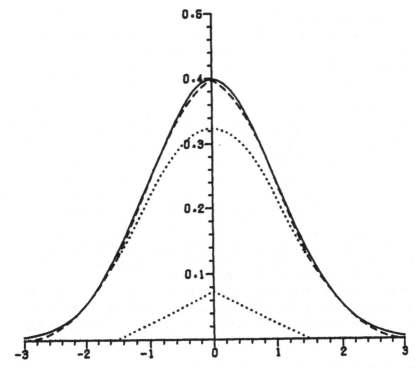

Figure 3.8. The Marsaglia–Bray composition algorithm for the normal distribution. The solid line is the normal pdf, the dotted lines are $p_1 f_1$ and $p_2 f_2$, and the dashed line is $p_1 f_1 + p_2 f_2$.

If $U < 0.8638$ then
   generate U_1, U_2, U_3,
   let $X = 2(U_1 + U_2 + U_3) - 3$
else if $U < 0.9745$ then
   generate U_1, U_2
   let $X = 1.5(U_1 + U_2 - 1)$
else if $U < 0.9973002039$
  repeat
    generate U_1, set $V = 6U_1 - 3$, generate U_2,
  until $0.358U_2 \leqslant g(V)$,
  let $X = V$
else
  repeat
    repeat
      generate U_1, U_2, set $V_i = 2U_i - 1$
    until $W = V_1^2 + V_2^2 < 1$,
    let $C = \sqrt{W^{-1}(9 - 2 \ln W)}$, $S = CV_1$, $T = CV_2$,

until $|S| > 3$ or $|T| > 3$
if $|S| > 3$ then $X = S$ else $X = T$
endif
Return X.

Here

$$g(x) = ae^{-x^2/2} - 2b(3 - x^2) - c(1.5 - |x|), \quad |x| < 1$$
$$ae^{-x^2/2} - b(3 - |x|)^2 - c(1.5 - |x|), \quad 1 \leqslant |x| < 1.5$$
$$ae^{-x^2/2} - b(3 - |x|)^2, \quad 1.5 \leqslant |x| < 3$$

$$a = 17.49731196, \quad b = 2.36785163, \quad c = 2.15787544$$

This algorithm uses on average around 3.9 uniforms per normal, but is usually faster than the simpler algorithms—see Table 3.2. If generating uniforms are slow, one can reuse U; see Exercise 3.10.

Marsaglia (1961a) and Marsaglia et al. (1964) use a 31-part composition for the positive part of the normal to which they attach a random sign. Figure 3.9 illustrates the idea. The pdfs $f_1, \ldots, f_{15}$ are uniform on specified

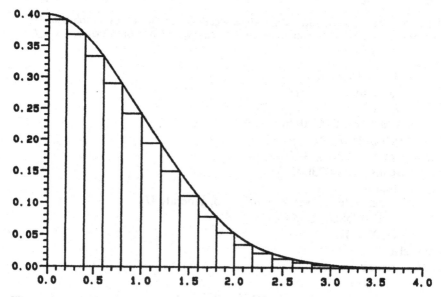

Figure 3.9. The rectangle-wedge-tail algorithm for the half-normal divides the area under the curve into the 31 regions shown.

intervals, and $f_{16}, \ldots, f_{30}$ are nearly triangular. Finally, f_{31} is the tail $X > 3$. The algorithm uses the following modified rejection technique for nearly linear densities.

Suppose $f = 0$ except on (a, b) and there are constants s and t such that

$$s - t(x - a)/(b - a) \leqslant f(x) \leqslant t - t(x - a)/(b - a)$$

on that interval. Then

1. Generate $U, V \sim U(0, 1)$.
2. If $U > V$ swap U and V.
3. If $V < s/t$ go to 5.
4. If $V > U + t^{-1}f(a + (b - a)U)$ go to 1.
5. Return $X = a + (b - a)U$.

Exercise 3.16 verifies that this works.

Marsaglia et al. (1976) refine this algorithm further. The reader is referred to the original papers for full details.

Kinderman and Ramage (1976) give a five-part composition algorithm for which the most frequently sampled distribution is the sum of two uniforms, $2.216(U_1 + U_2 - 1)$. Again, the paper gives full details. Ahrens and Dieter (1972) have yet another five-part composition method.

Yet more methods have been published by Déak (1980, 1981) and Dieter and Ahrens (1973). Marsaglia (1964) simulates from the tail of a normal distribution.

Exponential Distribution

We have seen two algorithms—3.2 (inversion) and 3.7 (von Neumann). Inversion is easy and is generally used unless it is too slow. The von Neumann method is faster provided random-number generation is fast. The ratio-of-uniforms method with pretests is also competitive. Marsaglia (1961b) and Knuth (1981, p. 128) give other methods.

Cauchy Distribution

The ratio-of-uniforms algorithm 3.8 is almost always preferred to inversion.

Student's t-Distribution

An obvious algorithm is normal$/(\chi_v^2/v)$, where χ_v^2 is regarded as a special case of a gamma distribution. Direct methods are given by Kinderman et al. (1977), Kinderman and Monahan (1980), and Marsaglia (1980). (See also Exercise 3.20.)

Gamma Distribution

The gamma distribution has been one of the most intensively studied in the last decade, so the survey of Atkinson and Pearce (1976) is hopelessly out of date. We consider only unit scale parameter, with pdf

$$x^{\alpha-1}e^{-x}/\Gamma(\alpha) \quad \text{on} \quad (0, \infty), \quad \alpha > 0$$

If 2α is an integer we can use twice a $\chi^2_{\alpha/2}$ variate for which we saw efficient methods for small α in Section 3.1. Otherwise we will consider separately the cases $\alpha > 1$ for which the pdf is bounded and $\alpha < 1$ for which it is not.

For $\alpha < 1$ we have the following simple and fast algorithm (Ahrens and Dieter, 1974).

Algorithm 3.19

1. Generate U_0, U_1.
2. If $U_0 > e/(\alpha + e)$, go to 4.
3. Let $X = \{(\alpha + e)U_0/e\}^{1/\alpha}$. If $U_1 > e^{-X}$ go to 1 else go to 5.
4. Let $X = -\ln\{(\alpha + e)(1 - U_0)/\alpha e\}$. If $U_1 > X^{\alpha-1}$ go to 1.
5. Return X.

This partitions $(0, \infty)$ into $(0, 1)$ and $(1, \infty)$ and uses separate envelopes on each. Its correct operation is verified in Exercise 3.17. Best (1983) refines this by splitting at $Z(\alpha) < 1$. See also Exercise 3.22.

For $\alpha > 1$, Cheng and Feast (1979) give three related algorithms derived from the ratio-of-uniforms method. The region C_h for $h(x) = x^{\alpha-1}e^{-x}$ is

$$\{(u, v)|u, v \geq 0, u^2 \leq (v/u)^{\alpha-1}e^{-v/u}\} \subset [0, a] \times [0, b]$$

where $a = [(\alpha - 1)/e]^{(\alpha-1)/2}$ and $b = [(\alpha + 1)/e]^{(\alpha+1)/2}$. We can express the test as $2 \ln U \leq (\alpha - 1) \ln X - X$. We would generate U as aU_1, giving

$$2 \ln U_1 + (\alpha - 1)\ln[(\alpha - 1)/e] \leq (\alpha - 1) \ln X - X$$

or, for $c = 2/(\alpha - 1)$ and $W = X/(\alpha - 1)$,

$$c \ln U_1 - \ln W + W \leq 1$$

This gives

1. Generate U_1, $U_2 \sim U(0, 1)$
2. Let $W = dU_2/U_1$ for $d \geq b/(\alpha - 1)a$

3. If $c \ln U_1 - \ln W + W > 1$ go to 1
4. Return $X = (\alpha - 1)W$

as one implementation. We can pretest step 3 by

2a. If $cU_1 + W - W^{-1} \leqslant c + 2$, go to 4

Figure 3.10 shows that as α increases the acceptance region shrinks toward the diagonal $U_1 = U_2$. Cheng and Feast thus suggested bounding C_h by a parallelogram. With suitable bounds for the constants we get

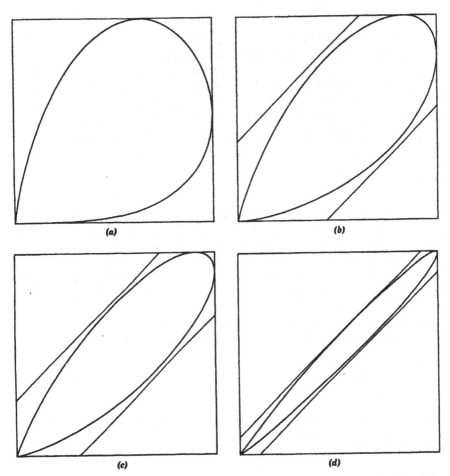

Figure 3.10. The region C_h in (U_1, U_2) space for the Cheng–Feast gamma algorithm 3.20. (a) $\alpha = 2$. (b) $\alpha = 5$. (c) $\alpha = 10$. (d) $\alpha = 100$. For $\alpha \geqslant 5$ the sloping lines are the intersection of the parallelogram with the unit square.

Algorithm 3.20 (gamma, $\alpha > 1$). $c_1 = \alpha - 1$, $c_2 = \{\alpha - 1/6\alpha\}/c_1$, $c_3 = 2/c_1$, $c_4 = c + 2$, $c_5 = 1/\sqrt{\alpha}$.

1.. Repeat
    generate U_1, U_2,
    if $\alpha > 2.5$ set $U_1 = U_2 + c_5(1 - 1.86U_1)$
    until $0 < U_1 < 1$.
2. Let $W = c_2 U_2 / U_1$.
3. If $c_3 U_1 + W + W^{-1} \leqslant c_4$ go to 5.
4. If $c_3 \ln U_1 - \ln W + W \geqslant 1$ go to 1.
5. Return $X = c_1 W$.

This uses a rectangle to enclose C_h for $1 < \alpha \leqslant 2.5$, and a parallelogram for $\alpha > 2.5$ to give a fairly constant speed as α is varied.

Many other algorithms have been proposed, including those of Ahrens and Dieter (1974, 1982b), Atkinson (1977), Atkinson and Pearce (1976), Best (1978, 1983), Cheng (1977), Cheng and Feast (1980a), Jöhnk (1964), Kinderman and Monahan (1980), Marsaglia (1977), Schmeiser and Lal (1980), and Tadikamalla (1978). Tadikamalla and Johnson (1981) give a survey at that date.

Beta Distribution

The beta and gamma distributions are related. If $X \sim$ gamma (α) and $Y \sim$ gamma (β), then $X/(X + Y) \sim$ beta(α, β). Conversely, if $X \sim$ beta $(\alpha, 1 - \alpha)$ and $Y \sim$ exponential then $YZ \sim$ gamma (α) for $\alpha < 1$. (This is Jöhnk's algorithm for the gamma.) For α and β both integers, the αth largest out of $U_1, \ldots, U_{\alpha+\beta+1}$ has a beta (α, β) distribution. Finally, beta variates can also be used to generate F-variates, for if $X \sim$ beta $(\tfrac{1}{2}\mu, \tfrac{1}{2}\nu)$, $\nu X/[\mu(1 - X)] \sim F_{\mu,\nu}$.

Many methods have been proposed for the beta distribution. Either of α or β can be less than one. The three main cases are

(i) $\alpha, \beta > 1$
(ii) $\alpha < 1 < \beta$
(iii) $\alpha, \beta < 1$

with $\beta < 1 < \alpha$ coming by symmetry ($X \to 1 - X$, $\alpha \leftrightarrow \beta$) and cases with $\alpha = 1$ or $\beta = 1$ by inversion.
Jöhnk (1964) gave

Algorithm 3.21 (beta $\alpha > 0$, $\beta > 0$)

Repeat
    Generate U_1, $U_2 \sim U(0, 1)$,

let $V_1 = U_1^{1/\alpha}$, $V_2 = U_2^{1/\beta}$
until $W = V_1 + V_2 \leqslant 1$.
Return $X = V_1/W$.

Exercise 3.18 verifies this method, and shows that the acceptance rate is poor if $\alpha + \beta$ is large. Ahrens and Dieter (1974), Atkinson (1979c), Atkinson and Pearce (1976), Atkinson and Whittaker (1976), Boswell and De Angelis (1981), Cheng (1978a), Sakesegawa (1983), Schmeiser and Babu (1980), and Schmeiser and Shalaby (1980) give faster and more complicated methods which apply in one or more of the three cases given previously. One can of course also use the relationship to the gamma, the order statistic method (for integer α and β), and inversion (if a suitable approximation is available).

Symmetric Stable Distributions

This is the family of distributions with characteristic functions $\exp\{i\mu t - |ct|^\alpha\}$. The cases $\alpha = 1$, the Cauchy, and $\alpha = 2$, the normal, are dealt with above. Chambers et al. (1976) show that if $V \sim U(-\pi/2, \pi/2)$ and $W \sim$ exponential then

$$\mu + c\frac{\sin(\alpha V)}{[\cos(V)]^{1/\alpha}}\left[\frac{\cos\{1 - \alpha)V\}}{W}\right]^{(1-\alpha)/\alpha}$$

has the required distribution.

More generally one can consider infinitely divisible distributions, of which stable distributions are a subclass. Bondesson (1982) uses an approximate composition method, letting X be a sum of random variables with randomly chosen distributions. For the stable distributions this reduces to the method of Bartels (1978).

3.5. RECOMMENDATIONS

It is impossible to make recommendations of which algorithm to use that are good across all environments. The first piece of advice is to follow Jackson's (1975) Rules for Programmers on optimization:

1. Don't do it.
2. (for experts only) Don't do it yet.

In other words, if you have a program available to you which will do what you want, use it. It is very unlikely that any lack of speed on its part will compensate for your time in programming and testing a preferred method.

That advice will apply to most mainframe installations but not to users of mini- and microcomputers and programmable calculators. For such

Table 3.3. Recomendations for Simple Algorithms

Normal:	Polar (3.6) or ratio (3.17).
Exponential:	Inversion (3.2).
Cauchy:	Ratio (3.8).
Student's t:	Normal/gamma or Exercise 3.20.
F:	Gamma/gamma or via beta.
Gamma:	$\alpha > 1$ Ahrens and Dieter (3.19) or Exercise 3.22.
	$\alpha > 1$ Cheng and Feast (3.20).
Beta:	$gamma_1/(gamma_1 + gamma_2)$ if gamma available, otherwise Jöhnk (3.21).
Geometric:	Integer part of exponential.
Binomial:	$n \leqslant 50$ count U_i's $\leqslant p$.
	$n > 50$ general method.
Poisson:	$\mu \leqslant 5$–20 multiplication (3.3).
	$\mu \leqslant 100$ general method.
	$\mu > 100$ algorithm 3.15.
Other discrete:	Indexed search with $m = \frac{1}{2}M$ or alias.

users, Table 3.3 recommends some simple algorithms that are unlikely to be embarrassingly slow in a simulation study.

There will occasionally arise the need to sample from other continuous distributions, when the methodology of Section 3.2 will be useful. Atkinson (1982) is one example, Tadikamalla (1979) another.

We have only considered the sensitivity of algorithms to the pseudo-random number generator used for certain methods for the normal distribution. The general experience seems to be that the algorithms described here are satisfactory when used with a good generator. However, it is always advisable to check that the final program does produce approximately the correct distribution, for example by a Kolmogorov–Smirnov test of the empirical cdf based on a large number (say 10,000) of samples.

Table 3.2 gives some comparative timings in a variety of environments. These times should be seen as a guide only, but confirm that relative speed varies little between very different computing environments.

EXERCISES

3.1.   Find a recursive formula for the pdf of the sum of n independent $U(0, 1)$ variates. [For the pdf note that its value on the interval $(i - 1, i)$, $i = 1, \ldots, n$ is a polynomial of degree $n - 1$ and obtain a recursion for the coefficients of these polynomials.] Hence show the pdf is

$$\sum_0^{\text{int}(x)} (-1)^r \binom{n}{r}(x - r)^{n-1}/(n - 1)! \quad \text{on} \quad (0, n)$$

3.2. Prove carefully that the Box–Muller algorithm returns independent standard normal variates. Show that if X and Y are independent with a differentiable pdf then R and Θ are independent only if X and Y are normally distributed.

3.3. Verify by direct calculation that if $N = \max\{n | E_1 + \cdots + E_n \leqslant t\}$ that N has a Poisson distribution, for independent exponential variates E_i.

3.4. Plot $\{(X, Y)\}$ for the Box–Muller and polar algorithms for the congruential generators $X_i = (aX_{i-1} + 1) \bmod 256, a = 133$ and 341.

3.5. Consider the beta density $f(x) = 6x(1 - x)$ on $(0, 1)$. Derive as many different algorithms as you can to sample from f, and compare them.

3.6. What is the expected number of uniforms used in the Forsythe method?

3.7. Verify that Algorithm 3.7 works. Derive the expected number of uniforms per exponential.

3.8. Apply the ratio-of-uniforms method to the t-distribution with pdf $\propto (1 + x^2/v)^{-(v+1)/2}$.

3.9. Hsuan (1979) gives a method for sampling from a polygonal region. Consider using it with the ratio method for normal and exponential distributions.

3.10. A frequent trick is to "reuse" uniforms. Show that if $U \sim U(0, 1)$ then $X = I(U \leqslant p)$ and $V = U/p$ if $X = 1$ otherwise $= (1 - U)/(1 - p)$ are independent, and $V \sim U(0, 1)$. Algorithm 3.19 uses this device.

3.11. Implement various methods for sampling from the distribution on $\{2, \ldots, 12\}$ of the sum of the outcomes of two dice throws.

3.12. For the alias method taking M a power of 2 allows Y to be found by shifts rather than multiplication. To do so one might introduce some zero probabilities. Do Algorithms 3.13 cope correctly? Try the example of Exercise 3.11.

3.13. Experiment with Algorithm 3.14 and choose K.

3.14. Show the McLeod and Bellhouse reservoir sampling method works in all cases.

3.15. Consider sampling without replacement with probability proportional to given weights. Can one do better than sampling with replacement and rejecting multiple occurrences? [See Wong and Easton (1980).]

3.16. Show that the modified rejection method for nearly linear densities works. What is its probability of acceptance?

3.17. Verify that Algorithm 3.19 works.

3.18. Verify that Algorithm 3.21 works.

3.19. Implement as many methods as you can for the triangular density $f(x) = \max(0, 1 - |x|)$.

3.20. Kinderman et al. (1977) give the following algorithm

   1. Generate $U, U_1 \sim U(0, 1)$.
   2. If $U < \frac{1}{2}$ then $X = 1/(4U - 1)$, $V = X^{-2}U_1$,
            else $X = 4U - 3$, $V = U_1$.
   3. If $V < 1 - \frac{1}{2}|X|$ go to 5.
   4. If $V \geq (1 + X^2/\nu)^{-(\nu + 1)/2}$ go to 1.
   5. Return X.

Show that this generates a t_ν-variate. [It is a rejection algorithm with $g(x) \propto \min(1, 1/x^2)$.]

3.21. Consider the inverse Gaussian distribution with pdf

$$(\lambda/2\pi x^3)^{1/2} \exp[-\lambda(x - \mu)^2/2\mu^2 x] \quad \text{on} \quad (0, \infty)$$

Show that $V = \lambda(X - \mu)^2/\mu^2 X \sim \chi_1^2$ and hence $X = X_1$ or X_2, where

$$X_1 = \mu + [\mu^2 V - \sqrt{\mu^2 V^2 + 4\mu\lambda V}]/2\lambda, \qquad X_2 = \mu^2/X_1$$

Show that we should choose X_1 with probability $\mu/(\mu + X_1)$ to obtain an inverse Gaussian variate (Michael et al., 1976). [Hint. Transform $X \to Y = \min(X, \mu^2/X) \to V = \lambda(Y - \mu)^2/\mu^2 Y$.]

3.22. Let Y have a gamma$(\alpha + 1)$ distribution, independent of U. Prove that $YU^{1/\alpha}$ has a gamma(α) distribution. This is an easy and fast method for $\alpha < 1$. On some systems it will be faster to generate $U^{1/\alpha}$ as $\exp(-E/\alpha)$ where $E \sim \exp(1)$.

3.23. Testing algorithms can be difficult. At some time line 3 of algorithm NMB of appendix B.5 was deleted, so only the first component of the Marsaglia–Bray algorithm 3.18 was used. How many samples would be needed to detect this error, say by a Kolmogorov–Smirnov or chi-square test?

CHAPTER 4

Stochastic Models

In Chapter 3 we were concerned almost exclusively with generating *independent* samples from specified distributions. In the simplest case this univariate distribution *is* the model, but often the stochastic model will utilize several dependent random variables. Constructing realizations with structured dependence has attracted much less attention. Many stochastic models can be simulated in the "obvious" way from their definitions, for example renewal processes. Nevertheless some tricks can be helpful and are presented below.

4.1. ORDER STATISTICS

Order statistics present perhaps the simplest form of dependence. An independent sample $(X_1, \ldots, X_n)$ is rearranged into increasing order as $X_{(1)} < \cdots < X_{(n)}$. The obvious way to do this is to sort the sample after generation. The fastest *general-purpose* sorting algorithms need $O(n \ln n)$ time to sort n items (Knuth, 1973b), so there will be a value of n for which the cost of sorting will dominate the cost of sampling. Typically this occurs for n in the range 100–10,000.

The alternative is to generate the sample in order. By taking $X_{(l)} = F^-(U_{(l)})$ we can reduce the problem to that of generating an ordered sample of random numbers, although in practice forming F^- may be very slow. Two methods have been suggested.

1. *Sequential.* Let $U_1, \ldots, U_n \sim U(0, 1)$. Define $U_{(n)} = U_n^{1/n}$, $U_{(k)} = U_{(k+1)} \times (U_k)^{1/k}, k = n - 1, n - 2, \ldots, 1$. This uses the fact that $U_{(1)}, \ldots, U_{(k)}$ are the order statistics of a sample of size k on $(0, U_{(k+1)})$, and plus inversion for $U_{(n)}$.

This method is particularly elegant for the exponential distribution. Let $E_{(k)} = -\ln U_{(n+1-k)}$. Then

$$E_{(k)} = E_{(k-1)} + W_k/(n + 1 - k)$$

where $E_{(0)} = 0$ and $W_1, \ldots, W_n$ are independent exponential variates.

96

2. *Exponential Spacings.* Let $E_1, \ldots, E_{n+1} \sim \exp(1)$ and $S_k = E_1 + \cdots + E_k$. Then $U_{(k)} = S_k / S_{n+1}$ have the required joint distribution (Exercise 4.1).

Another idea is to generate a partially sorted sample by *grouping.* Suppose the range of F is divided into intervals $I_1, \ldots, I_k$ of roughly equal probability. First a sample $(M_1, \ldots, M_k)$ of the multinomial distribution $(P(X \in I_i))$ is taken, then M_1 points from I_1 are sampled and sorted, followed by M_2 points from I_2, and so on. If k is chosen proportional to n, the average number of points sorted remains bounded and the expected time is $O(n)$. Rabinowitz and Berenson (1974) suggest k in the range $n/7$ to $n/2$ for the uniform distribution, and Gerontidis and Smith (1982) illustrate other distributions with $k \approx n/4$.

Consider the uniform distribution with $I_i = [(i - 1)/k, i/k)$. Then one way

Table 4.1. Timings for Methods for Generating All Order Statistics of a Sample of Size n^a

	n							
	10	20	100	1000	10	100	1000	10,000
	BBC Basic (msec)				VAX Fortran (μsec)			
Uniforms								
Sorting	16	23	32	46	100	125	170	220
Sequential	39	40	41	41	150	140	140	140
Exponential spacings	20	20	18	18	90	85	80	80
Grouping ($k \approx n/4$)	18	18	16	16	10	100	100	100
Normals								
Sorting	30	37	46	61	180	210	260	310
Inversion of $U_{(i)}$	55	53	52	52	240	225	225	220
Exponentials								
Sorting	28	32	44	57	135	170	210	260
Sequential	17	17	17	17	80	80	80	80
Inversion of exponential spacings	37	36	35	35	160	150	150	150

^{a}Times are per order statistic. In each case the fastest available algorithm was used for samples to be sorted.

to generate a multinomial sample is to generate n $U \sim U(0, 1)$ and count the number M_i of $U \in I_i$. Those $U_j \in I_i$ will do for the M_i points from I_i. Thus the method can be viewed as a special-purpose sorting method which uses our knowledge of the distribution of the numbers to be sorted. It is closely related to address-calculation and radix sorts (Knuth, 1973b, pp. 99–102, 177–178).

Table 4.1 gives some timings for the various methods. There is little to choose between them for moderate n; therefore if a good sort routine is available, it would seem the best choice. It is hard to envisage applications that would need all order statistics for $n \geqslant 1000$! Other comparisons are given by Lurie and Hartley (1972), Reeder (1972), Schucany (1972), Lurie and Mason (1973), Rabinowitz and Berenson (1974), Ramberg and Tadikamalla (1978), Bentley and Saxe (1980), and Gerontidis and Smith (1982).

The comparisons are rather different if we only want a small proportion of the order statistics, for example the median and quartiles. It is possible to modify sort algorithms to avoid sorting the whole sample (Chambers, 1977, pp. 45, 184) but the work of sampling remains. If we require just $X_{(k)}$ we can sample it as $F^{-}(U_{(k)})$, where $U_k \sim \text{beta}(k, n + 1 - k)$ can be sampled directly. (Note that in Chapter 3 we considered doing so via order statistics for small n.) If we want $X_{(k_1)}, \ldots, X_{(k_p)}, k_1 < \cdots < k_p$, we can adapt the sequential method, so $X_{(k_i)} = F^{-}(V_i)$, $V_i = V_{i+1} \times \text{beta}(k_i, k_{i+1} + 1 - k_i)$ variate, $i = p - 1, \ldots, 1$. Alternatively, we can modify the exponential spacings method, generating $W_i \sim \text{gamma}(k_i - k_{i-1})$ $i = 1, \ldots, p + 1$ ($k_0 = 0, k_{p+1} = n + 1$) and setting

$$X_{(k_i)} = (W_1 + \cdots + W_i)/(W_1 + \cdots + W_{p+1})$$

4.2. MULTIVARIATE DISTRIBUTIONS

The most common multivariate distribution is the multivariate normal in p dimensions with mean μ and dispersion matrix Σ. There are two general approaches to sampling $N(\mu, \Sigma)$.

Suppose $\Sigma = SS^T$ for some $p \times p$ matrix S. Then if $Z_1, \ldots, Z_p$ are independent normal deviates, $X = \mu + SZ \sim N(\mu, \Sigma)$ [since $E(X - \mu)(X - \mu)^T = SEZZ^TS^T = SS^T$]. Such matrices always exist. For example, the Cholesky decomposition of Σ is the unique lower-triangular matrix L with $LL^T = \Sigma$ (Nash, 1979). Then $X = \mu + LZ$ can be formed quite rapidly. This decomposition is particularly useful if Σ^{-1} is specified, for if $LL^T = \Sigma^{-1}$ then we can take $S = (L^{-1})^T$ and form $X = \mu + Y$, $L^TY = Z$. This triangular system of equations is easily solved.

The other general approach is to generate $Y \sim N(0, \Sigma)$ by generating Y_1, then Y_2 conditional on Y_1, and so on. Let A_k be the upper $k \times k$ submatrix of

Σ, and $\mathbf{a} = (\Sigma_{ik})_{1 \leqslant i \leqslant k-1}$. Then the conditional distribution of Y_k given $W_k = (Y_1, \ldots, Y_{k-1})^T$ is $N(\mathbf{a}^T A_{k-1}^{-1} W_k, \Sigma_{kk} - \mathbf{a}^T A_{k-1}^{-1} \mathbf{a})$. The vectors $\mathbf{a}^T A_{k-1}^{-1}$ and standard deviations $\sqrt{\Sigma_{kk} - \mathbf{a}^T A_{k-1}^{-1} \mathbf{a}}$ can be precomputed; A_k^{-1} can be found from A_{k-1}^{-1} without full inversion.

The implementation of the conditioning and Cholesky methods differs only in that

$$Y_k = \alpha_k Z_k + \beta_{1k} Z_1 + \cdots + \beta_{k-1,k} Z_{k-1}$$

for the Cholesky method and

$$Y_k = \gamma_k Z_k + \delta_{1k} Y_1 + \cdots + \delta_{k-1,k} Y_k$$

in the conditioning method. Thus both require the same time per X, but in general the Cholesky method will take less time to set-up. However, examples exist in which one form is easy to derive analytically. (In time series they are MA and AR representations respectively.)

Comparative studies of these methods include Scheur and Stoller (1962), Barr and Slezak (1972), and Hurst and Knop (1972). Generally the Cholesky method is preferred.

Wishart Distribution

Let $\mathbf{X}_k \sim N(\boldsymbol{\mu}_k, \Sigma)$, $k = 1, \ldots, n$ be independent multivariate normals with a common dispersion matrix. Let

$$W = \sum_1^n \mathbf{X}_k \mathbf{X}_k^T$$

Then W has a noncentral Wishart distribution $W(n, \Sigma, \Delta)$, where $\Delta = \sum \boldsymbol{\mu}_i \boldsymbol{\mu}_i^T$. It arises when considering sample covariance matrices.

Consider first the central case ($\Delta = 0$). Then $W = LVL^T$ where $LL^T = \Sigma$ and $V = \sum \mathbf{Z}_k \mathbf{Z}_k^T$, so a direct way to simulate V involves np normal deviates. We can reduce this load by Bartlett's decomposition. Let $\zeta_i \sim \chi^2_{n+1-i}$, $i = 1, \ldots, p$, and $\varepsilon_{ij} \sim N(0, 1)$ for $1 \leqslant i \leqslant j \leqslant p$. Then (Exercise 4.3)

$$V_{ii} = \zeta_i + \sum_{k<i} \varepsilon_{ki}^2$$

$$V_{ij} = V_{ji} = \varepsilon_{ij}\sqrt{\zeta_i} + \sum_{k<i} \varepsilon_{ki}\varepsilon_{kj} \quad \text{for } i < j$$

This algorithm is given by Odell and Feiveson (1966) and Chambers (1970). Smith and Hocking (1972) give Fortran code; they use an approximate method to generate the χ^2 variates whereas it would be better to use Algorithm 3.20.

The noncentral Wishart distribution depends only on Δ, a nonnegative definite symmetric matrix. Let $D = \Gamma\Gamma^T$ be its Cholesky decomposition. Take $\mu_1, \ldots, \mu_p$ as the rows of Γ, and $\mu_i = 0$ for $i > p$. Then

$$W = \sum_1^p (\mu_k + LZ_k)(\mu_k + LZ_k)^T + \sum_{p+1}^k LZ_k Z_k^T L^T$$

where the two terms are independent and the second has a $W(n - p, \Sigma, 0)$ distribution. The first term is $L[\sum (v_k + Z_k)(v_k + Z_k)^T]L^T$, where $Lv_k = \mu_k$, and can be simulated using p^2 normals. This decomposition follows Chambers (1970). Gleser (1976) gives an extended decomposition, which is both more complex and more efficient. See also Johnson and Hegemann (1974).

Discrete Distributions

In theory, multivariate discrete distributions are no different from any other discrete distributions; they still take a countable set of values. In practice we have to approximate this by a rather large finite set, so they provide a severe test of the standard methods. Bivariate distributions have been considered by Kemp and Loukas (1978, 1981) and Kemp (1981), and trivariate ones by Loukas and Kemp (1983). Straight searching is prohibitively slow, but, provided space is available, indexed, alias, and table methods performed well. One simple idea for a bivariate distribution (X, Y) on $\{1, \ldots, L\} \times \{1, \ldots, M\}$ with $P_{xy} = P(X < x) + P(X = x, Y \leq y)$ is to use the value of X as an index. That is, we first search on (P_{xM}) for $P(X \leq x - 1) < U \leq P(X \leq x) = P_{xM}$ and then search from P_{x0} to P_{xM} for y such that $P_{x,y-1} < U \leq P_{xy}$.

4.3. POISSON PROCESSES AND LIFETIMES

A point process in time is a sequence of times of occurrence $S_0 = 0, S_1, S_2, \ldots$. (We consider more general point processes in Section 4.6.) Let $T_i = (S_i - S_{i-1})$ be the times between occurrences, which we will think of as lifetimes. For a renewal process the T_i are independent, and the "obvious" way to simulate a renewal process is to construct the partial sums S_i of a sequence of independent samples T_i from the specified lifetime distribution.

A Poisson process of rate λ is a renewal process with $\exp(\lambda)$ lifetimes. This gives one way to simulate it. An alternative characterization is that the number N_t of points in $(0, t)$ has a Poisson (λt) distribution, and the N_t points are uniformly distributed on $(0, t)$. (If we want the points in order we could generate an ordered sample by the methods of Section 4.1; using exponential spacings then nearly recovers the renewal-process approach.)

Consider a heterogeneous Poisson process with rate function $\lambda(t)$. Define the cumulative rate $\Lambda(t) = \int_0^t \lambda(s)ds$. There are two immediate ways to simulate such a process. First, sample $N_t \sim$ Poisson$[\Lambda(t)]$ and then N_t points independently with cdf $\Lambda/\Lambda(t)$. Alternatively simulate a Poisson process S'_n of unit rate. Then $(S_n = \Lambda^{-1}(S'_n))$ is a heterogeneous Poisson process of rate $\lambda(t)$. This time-change transformation is an analogue of inversion and will generally be more efficient. [Kaminsky and Rumpf (1977) compare this with various approximations that are sometimes used.]

There is also an analogue of rejection sampling. Rejecting points in a Poisson process is known as *thinning*, an allusion to two-dimensional processes as models for forests. Lewis and Shedler (1979a) proposed the following algorithm.

Algorithm 4.1 (thinning).   Let (S_n) be a heterogeneous Poisson process of rate function λ, and h a function from $[0, \infty)$ to $[0, 1]$.

```
J = 0
for I = 1 to n
   repeat
      J = J + 1
      generate U ~ U(0, 1)
   until U ≤ h(S_J)
   S'_I = S_J
next I.
```

Then (S'_n) is a heterogeneous Poisson process of rate $h\lambda$.

Each point of (S_n) is retained independently with probability $h(S_n)$. Thus (S'_n) has independent numbers of points in disjoint intervals and so is a Poisson process. To find its rate function, note that

$$P(\text{some } S'_n \in (t, t + \Delta t)) \approx h(t)P(\text{some } S_n \in (t, t + \Delta t))$$

$$\approx h(t)\lambda(t)\Delta t$$

so the rate function is $h \times \lambda$.

A further operation for point processes is superposition, in which the points of two or more independent point processes are combined. The superposition of Poisson processes is a Poisson process with rate function the sum of those of the components. This can be used as an analogue of the composition principle to split λ into simpler components before the use of inversion or thinning.

Lewis and Shedler (1976, 1979b) illustrate these techniques for rate functions of the form $\exp(a_0 + a_1 t + a_2 t^2)$.

Lifetime Distributions

There is a close connection between lifetime distributions and heterogeneous Poisson processes. Lifetime distributions are often specified by their hazard function $h(t) = f(t)/[1 - F(t)]$, so

$$h(t)\Delta t \approx P(t \leqslant T < t + \Delta t | t \leqslant T)$$

Let $H(t) = \int_0^t h(s)ds$ be the cumulative hazard. Then $F(t) = 1 - \exp[-H(t)]$, and $F^-(u) = H^-[-\ln(1-u)]$. If $U \sim U(0, 1)$, $E = -\ln(1 - U) \sim \exp(1)$, so the inversion method for a distribution specified by its hazard is

$$T = H^-(E), \qquad E \sim \exp(1)$$

Now suppose (S_n) is a Poisson process with rate function h. Then S_1 is a sample from the distribution with hazard function h, for

$$P(S_1 > t) = P(N_t = 0) = \exp[-H(t)] = 1 - F(t)$$

Hence we can use any of the methods described for heterogeneous Poisson processes to simulate lifetime distributions. We have just considered inversion. Superposition leads to

$$T = \min(T_1,\ldots, T_k), \qquad \text{haz}(T_l) = \lambda_l, \qquad \lambda_1 + \cdots + \lambda_k = h$$

and hence tends to be slow. Thinning gives

Algorithm 4.2. Suppose $h \leqslant g$, a hazard function with cumulative hazard G.

```
S = 0
Repeat
  generate E ~ exp(1), U ~ U(0, 1)
  let S = S + E, Y = G⁻(S)
until Ug(Y) ≤ h(Y).
Return T = Y.
```

This is just $I = 1$ in Algorithm 4.1 applied to the Poisson process of rate g simulated by inversion. Let N denote the number of exponentials or uniforms generated. Then (Devroye, 1985)

$$EN = \int_0^\infty f(t)G(t)dt$$

To see this, let S_i be the successive points tried. Then (S_i) is a Poisson process of rate g, and $(S_i, Ug(S_i))$ form a homogeneous Poisson process on $\{(t, x)|0 \leqslant x \leqslant g(t)\}$. Thus if $T > t$ we expect to reject

$$\int_0^t [g(u) - h(u)]du = G(t) - H(t)$$

points by time t. From this

$$EN = 1 + E[G(T) - H(T)] = 1 + EG(T) - 1$$
$$= \int_0^x G(t)f(t)dt$$

Alternative expressions are

$$EN = \int_0^x g(t)[1 - F(t)]dt = \int_0^x g(t)f(t)/h(t)dt \leqslant \sup\left(\frac{g(t)}{h(t)}\right)$$

confirming that if g and h match well then Algorithm 4.2 is efficient.

Devroye (1985) has an elegant modification for densities with decreasing hazard rates, which he terms "dynamic thinning."

Algorithm 4.3. Suppose $h(0) < \infty$ and h is nonincreasing.

    $T = 0$
    Repeat
      bound $= h(T)$
      generate $E \sim$ exp(bound), $U \sim U(0, 1)$
      let $T = T + E$
    until $U \times$ bound $\leqslant h(T)$.
    Return T.

This corresponds to using a point process with rate function $h(S_n)$ on $[S_n, S_{n+1})$, which is no longer a Poisson process. However,

$$P(t \leqslant T < t + \Delta t | t \leqslant T) \approx P(t \leqslant E + T_0 < t + \Delta t | t \leqslant E + T_0) \times [h(t)/\text{bound}]$$
$$= (1 - e^{-\text{bound}\Delta t})h(t)/\text{bound} \approx h(t)\Delta t$$

Table 4.2. Timings for the Pareto Distribution[a]

	α					
	0.1	0.5	1	5	10	100
Inversion	21	8	6	38	38	38
Algorithm 4.2	∞[b]	∞[b]	∞[b]	32	30	28
Algorithm 4.3	135	60	47	35	31	30

[a]Milliseconds per random variable in BBC Basic.
[b]Infinite theoretical mean and large unstable empirical values.

where T_0 is the largest value of T less then t. Thus T has hazard function $h(\ )$ as required.

It is not possible to give a general formula for EN for Algorithm 4.3. Devroye (1985) gives a number of bounds.

Example. The Pareto distribution with $f(t) = \alpha/(1 + t)^{1+\alpha}$ on $(0, \infty)$, with $\alpha > 0$. Then $h(t) = \alpha/(1 + t)$ and $H(t) = \alpha \ln(1 + t)$, so inversion yields

$$T = \exp(E/\alpha) - 1 = U^{-1/\alpha} - 1$$

for $E \sim \exp(1)$ or $U \sim U(0, 1)$. The dynamic thinning method can also be used. Devroye (1985) shows

$$EN = 1 \bigg/ \int_0^\infty e^{-z} \left(1 + \frac{z}{\alpha}\right)^{-1} dz \leqslant 1 + \alpha^{-1}$$

For constant hazard $g(t) \equiv \alpha$ we find $EN = \alpha ET = \alpha/(\alpha - 1)$ for $\alpha > 1$ and ∞ for $\alpha \leqslant 1$. Despite this, Table 4.2 shows little advantage for dynamic thinning.

For an *ordered* sample of lifetime distributions the obvious methods are to sort a sample or to use $H^-(E_{(i)})$, where $E_{(i)}$ are generated by the sequential method (Newby, 1979).

4.4. MARKOV PROCESSES

Many of the stochastic models of operations research can be regarded as Markov processes, and Markov models are also used in demography and population biology. A discrete-time Markov chain with state space S and transition matrix P can be simulated in the obvious way:

> If $X_n = s$ select X_{n+1} from the
> discrete distribution $\{p_{si} \mid i \in S\}$.

It may be worth setting up index search or alias tables for the rows of P, or at least those more commonly visited if S is large. One could, for example, set up alias tables for row s on the first visit to s.

An alternative is to compute T_n, the time to the next jump, then X_{n+T_n}. The wait T_n has a geometric distribution with parameter p_{ss} if $X_n = S$, and X_{n+T_n} has the discrete distribution $\{p_{si}/(1 - p_{ss}) \mid i \in S \setminus \{s\}\}$. This is likely to be faster at the expense of program complexity.

Continuous-time Markov chains can be simulated in the same way. Let τ_0, $\tau_1, \ldots$ be the sequence of times between jumps, and $J_0, J_1, \ldots$ be the sequence of states visited. Then we can reconstruct the process from (J_n, τ_n) (technically only up to the "first infinity"). Furthermore, conditional on $(J_0, \ldots, J_n, \tau_0, \ldots, \tau_{n-1})$, τ_n is exponentially distributed with rate q_{J_n}, and (J_n) is a Markov chain. Thus we can simulate the continuous-time process by sampling an exponential time to the next transition, then sampling the next state from the J_nth row of the transition matrix J of (J_r). Here $J_{rs} = q_{rs}/q_r$ and $J_{rr} = 0$. A formal version of this construction is given by Freedman (1971, Theorem 7.33).

Looking at a process in this fashion is known as *discrete-event* simulation. It may well be possible to obtain all the information we need from (J_n) without knowing (τ_n). For example, in the queueing model for a bank discussed in Chapter 1 we might suppose that a customer leaves in disgust if the queue to which he or she is assigned already contains 10 or more customers. Then we can find from (J_n) the proportion of customers who leave. It is also true for a wide class of queueing models that the equilibrium distribution of the process does not depend on the distribution of (τ_n) (Kelly, 1979), so we may as well take τ_n to be discrete.

Although many models are Markov or have embedded Markov chains (a series of times at which observations form a Markov chain), this is not the most natural way to view a queueing system. The system is specified by what happens to customers and servers, and discrete-event simulation languages such as SIMSCRIPT, GPSS, and Simula work with descriptions in these terms. They are described and compared in Fishman (1978a) and Bratley et al. ((1983). Hordijk et al. (1976) discuss some of the theory.

4.5. GAUSSIAN PROCESSES

Gaussian processes on $T \subset \mathbf{R}^d$ are stochastic processes all of whose joint distributions are jointly normal. That is, a random variable $X(t)$ is defined for each $t \in T$, and $X(t_1), \ldots, X(t_r)$ have a multivariate normal distribution for each r-tuple $(t_1, \ldots, t_r)$ and each r.

One will only need to sample a stochastic process at a finite number of points, so a sample from a Gaussian process is just one from a multivariate normal distribution. However, the number of points can easily be 1000 or 10,000, in which case one faces the decomposition of an $n \times n$ matrix for n of the order of 10^3 or 10^4. This is computationally prohibitive. More efficient methods can be found using stationarity.

Consider first time series models, with $T = \{1, 2, 3, \ldots\}$. An $MA(q)$ process is defined by

$$X_t = \varepsilon_t + \beta_1 \varepsilon_{t-1} + \cdots + \beta_q \varepsilon_{t-q} \tag{1}$$

where $\varepsilon_t \sim N(0, \sigma^2)$ independently for $t \in \{-q, -q+1, \ldots\}$. This can be simulated straight from the definition (1). An $AR(p)$ model has

$$X_t = \alpha_1 X_{t-1} + \cdots + \alpha_p X_{t-p} + \varepsilon_t \tag{2}$$

In this case we need to specify $X_1, \ldots, X_p$ to start the recursion. Since this is a Gaussian process, $(X_1, \ldots, X_p) \sim N(0, \Sigma)$ and we only need to find the dispersion matrix Σ. This is done using stationarity, for Σ is the dispersion matrix of

$$\begin{bmatrix} X_t \\ \cdot \\ \cdot \\ \cdot \\ \cdot \\ X_{t+1-p} \end{bmatrix} = \begin{bmatrix} \alpha_1 & \cdots & \cdot & \alpha_p \\ 1 & & & 0 \\ \cdot & & & \cdot \\ \cdot & & & \cdot \\ 0 & \cdots & 1 & 0 \end{bmatrix} \begin{bmatrix} X_{t-1} \\ \cdot \\ \cdot \\ \cdot \\ X_{t-p} \end{bmatrix} + \varepsilon_t \tag{3}$$

so

$$\Sigma = A\Sigma A^T + \begin{bmatrix} \sigma^2 & \cdots & 0 \\ \cdot & & \cdot \\ \cdot & & \cdot \\ 0 & \cdots & 0 \end{bmatrix}$$

where A denotes the $p \times p$ matrix in (3). This system is solved to find the symmetric matrix Σ. [See, for example, Gardner et al. (1979).] Alternatively one could set $X_1 = \cdots = X_p = 0$ and run (2) for long enough to settle down to equilibrium. This can be quantified, for the effect of the initial conditions decays as θ^{-t}, where θ is the smallest modulus of a root of

$z^p - \alpha_1 z^{p-1} - \cdots - \alpha_p$. By combining the ideas for AR and MA processes one can simulate $ARMA$ processes.

The continuous-time analogues of $ARMA$ models (linear stochastic difference equations) are stochastic differential equations and Brownian motion. Like ordinary differential equations these can be solved by discretization into small time steps, so the increments of Brownian motion become independent normal variates, and Brownian motion itself is approximated by a random walk [see Rao et al. (1974) and Platen (1981)]. There is one useful dodge that can be used to reduce the time step where necessary. [For example, Knuth (1984).] Suppose $B(t)$ is Brownian motion with $\mathrm{var}[B(t)] = \sigma^2 t$, and $B(n\tau)$ and $B(n\tau + \tau)$ have been found. Define

$$Y(s) = [B(n\tau + s\tau) - B(n\tau)] - s[B(n\tau + \tau) - B(n\tau)], \qquad 0 \leqslant s \leqslant 1$$

Then knowledge of the Y process would allow $B(t)$ to be filled in on $(n\tau, n\tau + \tau)$. The process Y is a Brownian bridge. To simulate a Brownian bridge one can reverse the process; let $W(t), 0 \leqslant t \leqslant 1$, be a random-walk approximation to Brownian motion and let $W_0(t) = c[W(t) - tW(1)]$. Then W_0 is a Brownian bridge (Exercise 4.7).

A Gaussian process on $\mathbb{R}^d$ is specified by its mean function $m(\ )$ (which we will assume to be zero) and covariance function $C(x, y)$. Under the assumption of stationarity under *translations* we have $C(x, y) = C(x - y, 0) = c(x - y)$, say. If in addition the process is stationary under rotations (*isotropic*), $c(h)$ is a function of the length of h only, so $C(x, y)$ only depends on the distance between x and y. There are few direct ways of simulating a Gaussian process with a specified covariance function. If we can simulate *any* process with covariance function C, we can obtain an approximation to a Gaussian process with the same covariance function by averaging a number of realizations of the non-Gaussian process and relying on the central limit theorem.

Zubrzycki (1957) took a Poisson process of rate λ on $\mathbb{R}^d$ and centered a disc of radius R about each point. Then $Z(x)$ is the number of discs that cover x; it has an isotropic covariance function which decays to zero at $2R$. Sironvalle (1980) allowed R to be random, chosen independently for each disc from a distribution F. On $\mathbb{R}^2$

$$c(r) = \lambda \int_r^\infty \left[\frac{x^2}{2} \cos^{-1}\left(\frac{r}{x}\right) - \frac{r}{2}(x^2 - r^2)^{1/2} \right] dF(x)$$

so

$$c''(r) = \lambda \int_r^\infty \frac{r}{\sqrt{x^2 - r^2}} dF(x)$$

and

$$1 - F(r) = \frac{2}{\pi\lambda}\int_r^\infty \frac{c''(s)ds}{\sqrt{s^2 - r^2}}$$

This will *not* give a distribution function for arbitrary c, but will do so for many covariance functions if λ is chosen appropriately. This clearly gives a non-Gaussian process, but averaging independent copies amounts to taking λ large and considering $\lambda^{-1}X_\lambda$.

Another approximate method for isotropic covariances is the "turning band" method of Matheron (1973). Simulate a stationary process Z_1 on $\mathbb{R}$ with covariance function c_1. Select a random rotation O in $\mathbb{R}^d$, and define $Z(x) = Z_1((Ox)_1)$. This is a stationary isotropic non-Gaussian process with covariance function

$$c(r) = \frac{2\Gamma(d/2)}{\sqrt{\pi}\Gamma[\frac{1}{2}(d-1)]}\int_0^1 c_1(vr)(1 - v^2)^{(d-3)/2}dv \tag{4}$$

We need to invert (4) to find c_1 from c. [There is always a suitable c_1 (Ripley, 1981, pp. 12–13).] This is particularly simple for $d = 3$ when

$$c_1(r) = \frac{d}{dr}[rc(r)]$$

Again averaging is needed to produce an approximately Gaussian process.

The turning bands method needs samples of stationary processes on $\mathbb{R}^1$ and we have as yet seen few examples. Solutions of stochastic differential equations provide one possibility; another is to generalize moving average processes to

$$X(x) = \int k(x - y)dB(y) \tag{5}$$

or

$$X(x) = \int k(x - y)dN(y) = \sum k(x - y_i) \tag{6}$$

for Brownian motion B or a Poisson process N with points $\{y_i\}$. Both have the covariance function

$$c(x) = \int k(x - u)k(u)du \qquad (7)$$

but (6) has a nonzero mean and is non-Gaussian. The stochastic integral (5) will be approximated by a sum of the form

$$X(x) \approx \sum k(x - n\delta)\varepsilon_n$$

where $\varepsilon_n \sim N(0, \delta)$, independently for each n. This is a convolution and so can be evaluated rapidly at $\{x = r\delta\}$ via Fast Fourier Transforms.

To do so we regard $X_j = X(j\delta)$ as a stationary process on $\{0, \ldots, N - 1\}$ with addition mod N. Then if

$$X_j = \sum_{r=0}^{N-1} k(j\delta - r\delta)\varepsilon_r \qquad (8)$$

and we define $k_j = k(j\delta \bmod N\delta)$, then

$$X_j = \sum_s k_{j-s}\varepsilon_s$$

Let $\tilde{X}$ and $\tilde{\varepsilon}$ be discrete Fourier transforms of X and ε, so

$$\tilde{X}_r = \sum_{j=0}^{N-1} e^{2\pi i j r/N} X_j$$

$$= \sum_j e^{2\pi i j r/N} \sum_s k_{j-s}\varepsilon_s = \sum_s \varepsilon_s \sum_j e^{2\pi i j r/N} k_{j-s}$$

$$= \sum_s \varepsilon_s \sum_k e^{2\pi i r(s-k)} k_k = \tilde{k}_{-r}\tilde{\varepsilon}_r$$

We can obtain (X_j) from $(\tilde{X}_r)$ by

$$X_j = N^{-1} \sum e^{-2\pi i j r/N} \tilde{X}_r$$

This gives us a rapid way to evaluate (8). Form $(\tilde{k}_r)$ in advance. Generate $(\varepsilon_0, \ldots, \varepsilon_{N-1})$, form $(\tilde{\varepsilon}_r)$ then $(\tilde{X}_r)$ and (X_j). If N is a power of 2 the Fourier transforms can be calculated via the Fast Fourier Transform. In fact, we can generate $(\tilde{\varepsilon}_r)$ directly. Let

$$\tilde{\varepsilon}_r = U_r + iV_r$$

Then $V_0 = 0$, $U_{N-r} = U_r$ and $V_{N-r} = -V_r$, giving N linearly independent variables $(U_0, \ldots, U_{N/2}, V_1, \ldots, V_{N/2-1})$ provided N is even. These are also statistically independent, normally distributed with variance $N\delta/2$ except U_0, which has variance $N\delta$.

We have some freedom in this construction, for (7) does not determine $k(\ )$ completely. All we have is $\tilde{c} = |\tilde{k}|^2$. Thus we may without loss of generality take $\tilde{k}_r = \tilde{k}_{-r} = \sqrt{\tilde{c}}$. [Here $\tilde{c}$ is the spectral density of (X_j).] This gives the following algorithm.

Algorithm 4.4. Covariance function $c(r\delta)$ specified, $r = 0, \ldots, N/2 - 1$ for even N.

1. Form $c_0, \ldots, c_{N-1}$ by $c_r = c(r\delta)$, $r \leqslant N/2 - 1$ and $c_{N-r} = c_r$.
2. Form $\tilde{c}_s$ and $\phi_s = \sqrt{\tilde{c}_s}$, $s = 0, \ldots, N - 1$.

For each simulation of $X(0), \ldots, X((N - 1)\delta)$

1. Sample $U_0 \sim N(0, N\delta)$, $U_1, \ldots, U_{N/2} \sim N(0, N\delta/2)$, $V_1, \ldots, V_{N/2-1} \sim N(0, N\delta/2)$.
2. Let $U_{N-r} = U_r$, $r = 1, \ldots, N/2 - 1$.
   Let $V_0 = V_{N/2} = 0$, $V_{N-r} = -V_r$, $r = 1, \ldots, N/2 - 1$
3. For $j = 1, \ldots, N - 1$ let

$$X(j\delta) = N^{-1} \sum_0^{N-1} \{e^{2\pi ijr/N}(U_r + iV_r)\phi_r\}$$

via a Fast Fourier Transform algorithm.

This method will generate any stationary Gaussian process on $\{0, \ldots, N - 1\}$ and approximates any stationary Gaussian process on $\mathbb{R}$. The circularity will give a poor approximation unless $N\delta \gg$ range of c. Davis et al. (1981) give Fortran code to implement this algorithm. (Lines 1400 and 1670–1700 ensure $\overline{X} = 0$ and should be omitted, and a better normal generator used.)

4.6. POINT PROCESSES

Section 4.3 considered only Poisson point processes on $[0, \infty)$. Renewal processes can be simulated directly from their definition as partial sums of independent lifetimes. Sometimes we will want a renewal process in equilibrium (that is, starting observation at an arbitrary time, not at an event). Let T_1 be the time from beginning observation to the first event observed, and T_2, $T_3, \ldots$ subsequent inter-event times. Then T_1 has a different distribution from T_2, with density

$$d(t) = \frac{1}{\mu}[1 - F(t)] \quad \text{on} \quad (0, \infty)$$

where F is the cdf of T_2, and μ is its mean. Thus to simulate the process on $(0, \infty)$ we sample T_1 from d, then T_2, $T_3, \ldots$ from the normal lifetime distribution.

Another approach is necessary to simulate an equilibrium renewal process on $(-\infty, \infty)$. Clearly 0 will belong to a lifetime with probability proportional to the length of the interval. Thus the interval (T_{-1}, T_1) containing zero has a length with pdf $tf(t)/\mu$ and is uniformly distributed about zero. Once T_{-1} and T_1 are established the simulation is completed both forward and backward in the obvious way.

Most other models of point processes in time can be simulated directly from their definitions. [See, for example, Cox and Isham (1980).]

Stochastic geometry and spatial statistics (Ripley, 1981) make use of point processes on quite complicated spaces. Most specific models are based on the Poisson process. The only construction from Section 4.3 that can be generalized to points on the plane is that which distributes N_t points inependently on $(0, t)$. For very general spaces X we can define a Poisson process N with mean measure Λ by

(i) $N(A)$ is the number of points in the (measurable) subset $A \subset X$.

(ii) If $\{A_r\}$ are disjoint, $\{N(A_r)\}$ are independent.

(iii) $N(A) \sim$ Poisson $[\Lambda(A)]$.

(iv) Conditional on $N(A) = n$, the n points in A are independently distributed with distribution $\Lambda/\Lambda(A)$.

Some of these properties make sense only if $\Lambda(A)$ is finite. Since we will only be able to simulate a finite number of points, we will confine attention to $A \subset X$ with $\Lambda(A) < \infty$. The properties (iii) and (iv) give a direct way to simulate the process on A. It may be difficult, however, to sample from the multidimensional distribution given at (iv). In this case we could partition $A = A_1 \cup \cdots \cup A_r$, $A_i \cap A_j = \varnothing$, and use property (ii) to simulate independently on each A_i. Then Λ may be more nearly uniform on A_i and so easier to sample. Figure 4.1 shows a simple example. The thinning method also generalizes in an obvious way.

Poisson cluster processes are defined by replacing each point of a Poisson process with an independent group of points. Cox or doubly stochastic

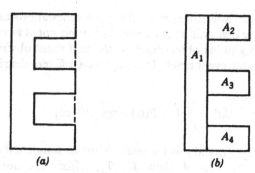

(a)   *(b)*

Figure 4.1. Subdividing a region A for a Poisson process. In *(a)* one samples points in the outer rectangle, for *(b)* for $A_1, \ldots, A_4$ in turn.

Poisson processes are Poisson conditional on the realization of the *random* mean Λ. Both may be simulated directly from their definitions.

Gibbsian point processes, which are often used as models for regular point patterns, pose more difficulty. Again we confine attention to a set $A \subset X$ for which $P(N(A) < \infty) = 1$. Then we can specify the distribution of the $N(A)$ points by the $\{P(N(A) = n)\}$ and the pdfs f_n of points $(x_1, \ldots, x_n)$, *conditional* on $N(A) = n$. Unfortunately f_n may be far from constant and of an intractable form. Consider, for example, n points in $\mathbb{R}^d$ distributed at random in the unit ball, *conditional on no two points being closer than r*. Rejection sampling is possible in theory but may be phenomenally inefficient (Ripley, 1977, p. 179). An obvious alternative is to generate x_1, then x_2 conditional on x_1, and so on. However, in our example we do not even know the marginal distribution of x_1. Fortunately f_n is known to be the equilibrium density of a space-time process that is easy to simulate (Ripley, 1977, 1979).

The density f_n cannot depend on the order of $x_1, \ldots, x_n$, since this order has no physical meaning. We just assume that $f(x_n | x_1, \ldots, x_{n-1})$ is known. Consider the following algorithm.

Algorithm 4.5. Start with $X_1, \ldots, X_n$ as any set of points with $f_n(X_1, \ldots, X_n) > 0$.

> For step $= 1$ to large number
> Generate U. Let $k = 1 + \text{int}[NU]$.
> Generate X from $f(X | X_1, \ldots, X_{k-1}, X_{k+1}, \ldots, X_n)$.
> Replace X_k by X.

It can then be shown under very mild conditions on f_n (for example, $f_n > 0$ suffices) that the limiting distribution of $(X_1, \ldots, X_n)$ has density f_n (Ripley, 1979; Lotwick and Silverman, 1981). How large the number of steps needs to be depends on f_n, but usually $10n$ steps will suffice to eliminate the initial conditions, and samples taken $2n$–$4n$ steps apart are virtually independent.

Algorithm 4.5 can be thought of as a spatial birth-and-death process in which deaths and births alternate. More general processes of this type cover the cases in which $P(N(A) = n)$ is unknown (Preston, 1977).

4.7. METROPOLIS' METHOD AND RANDOM FIELDS

The idea of using a Markov chain to simulate a complex system (as for a point process in Section 4.6) has been used extensively in statistical physics following the early work of Metropolis [in Metropolis et al. (1953)]. We will set out a general framework and illustrate it with the simulation of random fields.

Suppose we have a system with a large but finite number of states, and we wish to sample from a distribution that gives state j probability $\pi_j > 0$. A Markov chain with transition matrix P will have equilibrium distribution π if and only if

$$\pi^T = \pi^T P \tag{9}$$

and this chain is reversible if and only if the detailed balance conditions

$$\pi_i p_{ij} = \pi_j p_{ji} \qquad \text{for all } i \neq j \tag{10}$$

are satisfied (Kelly, 1979, Section 1.2). It is easy to see that (10) implies (9), for

$$(\pi^T P)_j = \sum_i \pi_i p_{ij} = \sum_i \pi_j p_{ji}$$

$$= \pi_j \sum_i p_{ji} = \pi_j$$

To sample from π we select a Markov chain with transition matrix P satisfying (10) and run it until it appears to have settled down to equilibrium. A general way to choose P is

Algorithm 4.6 (Metropolis).   Choose a *symmetric* transition matrix Q. At each step select j from Q_i., and move from i to j with probability $\min\{1, \pi_j/\pi_i\}$ otherwise remain at i.

This defines

$$p_{ij} = \min\{1, \pi_j/\pi_i\} q_{ij}$$

for $j \neq i$ and

$$p_{ii} = q_{ii} + \sum_{j \neq i} \max\{0, 1 - \pi_j/\pi_i\}q_{ij}$$

for which (10) holds, for

$$\pi_i p_{ij} = \min\{\pi_i, \pi_j\}q_{ij}$$
$$= \min\{\pi_i, \pi_j\}q_{ji} = \pi_j p_{ji}$$

There remain the questions of uniqueness and convergence to the unique equilibrium distribution. The equilibrium distribution will be unique if P is irreducible. However, $p_{ij} > 0$ if and only if $q_{ij} > 0$, so P is irreducible if and only if Q is. The distribution of (X_n) will converge to π if P is aperiodic. Note that $p_{ii} > 0$ unless both $q_{ii} = 0$ and $\pi_j \geqslant \pi_i$ for all states j with $q_{ij} > 0$. If for one of these states $\pi_j > \pi_i$, then $p_{jj} > 0$. Thus except in the trivial case of constant π_i, at least one $p_{jj} > 0$, so the state j and hence the whole chain is aperiodic. Thus a sufficient condition if π is not constant is to check that it is possible to move from any state to any other under Q.

Other prescriptions of P have been proposed. Barker (1965) replaced $\min\{1, \pi_j/\pi_i\}$ by the smaller quantity $\pi_j/(\pi_i + \pi_j)$. Again

$$\pi_i p_{ij} = \frac{\pi_i \pi_j}{\pi_i + \pi_j}q_{ij} = \pi_j p_{ji} \quad \text{for } i \neq j$$

so (10) holds. (In this case P is always aperiodic.) Hastings (1970) gives a parametric family of prescriptions that includes both Barker's and Metropolis' methods. Peskun (1973) shows this is inferior to Metropolis' suggestion for estimating $E_\pi f(X)$ by the average of $f(X_n)$. This says little, however, about the rate of convergence of X_n to equilibrium. (Metropolis' method is better than Barker's principally because it makes more transitions, so the average is over more independent terms.)

Markov Random Fields

Consider a process that takes one of k colors at each of a set of sites. These sites will be related by a graph structure, for example, sites on a square lattice, so each site has a number of neighbors. The process is specified by giving the joint distribution of all the random variables $(x_s | s \in S)$. A Markov random field has the special property that $P(x_r = \text{color} | x_s, s \neq r) = P(x_r = \text{color} | x_s, s$ a neighbor of $r)$ for each site. In words, the conditional distribution at a site given the rest depends only on the values at the neighboring sites.

Such processes are specified by giving these conditional distributions, with the result that the joint distribution is known only up to a normalizing constant (Kelly, 1979, Chapter 9).

As an example, consider a binary process on a $M \times M$ lattice with colors white and black. We can define

$$P(x_r = \text{black} \mid \text{rest}) = \frac{e^\eta}{1 + e^\eta}$$

where

$$\eta = \beta(\text{number of black neighbors} - \text{number of white neighbors})$$

This corresponds to the joint distribution

$$P(x_1 \cdots x_{MM}) = \text{const} \exp\{\beta \sum_{\substack{\text{neighbor} \\ \text{pairs of sites}}} 1(\text{same color})\} \tag{11}$$

but the constant cannot be written in any simple form.

This is an ideal problem for Markov chain methods, which only depend on π_j / π_i. We can choose Q so that $q_{ij} > 0$ only if i and j are states that differ in value at a single site. If this site is r, then

$$\frac{\pi_j}{\pi_i} = \frac{P(x_r = \text{new color} \mid \text{rest})}{P(x_r = \text{old color} \mid \text{rest})}$$

which is easily computed for a Markov random field. For example, we might select Q by choosing a site at random and changing its value to a randomly chosen color (not necessarily excluding the current color). This is clearly symmetric and irreducible so Algorithm 4.6 can be used.

Barker's method for a *binary* Markov random field can be expressed rather more simply. At the chosen site $x_r = \text{black}$ or white so $\pi_i + \pi_j = P(x_r = \text{white, rest}) + P(x_r = \text{black, rest}) = P(\text{rest})$ and

$$\frac{\pi_j}{\pi_i + \pi_j} = \frac{P(x_r = \text{new color, rest})}{P(\text{rest})} = P(x_r = \text{new color} \mid \text{rest})$$

The method thus reduces to choosing the new color at r from the conditional distribution given the neighboring values. This can also be considered for $k > 2$. Geman and Geman (1984) called this variant the "Gibbs sampler."

Yet another method based on Markov chains is Flinn's (1974) "spin exchange" method. Two sites are selected at random and the interchange of their values considered, being performed with probability π_j/π_i. Suppose r and s are the sites selected. Then

$$\frac{\pi_j}{\pi_i} = \frac{P(x_r = \text{old } x_s|\text{rest})P(x_s = \text{old } x_r|\text{rest})}{P(x_r = \text{old } x_r|\text{rest})P(x_s = \text{old } x_s|\text{rest})}$$

provided r and *s* are not neighbors. If *r* and *s are* neighbors, the conditional joint distribution of x_r and x_s must be evaluated. For example, from (11) we find

$$\frac{P(x_r = \text{white}, x_s = \text{black} \mid x_t, t \neq r, s)}{P(x_r = \text{black}, x_s = \text{white} \mid x_t, t \neq r, s)} = \exp \beta(W_r - B_r + B_s - W_s)$$

where W_r and B_r are the numbers of white and black neighbors of r excluding s, and W_s and B_s analogously for s excluding r. This method was used by Cross and Jain (1983) to simulate models of image textures (apparently without considering the neighbours problem). The Markov chain produced is only irreducible over the set of states with the same marginal distribution. For example, the proportion of black sites is unchanged. Thus the method produces a conditional simulation of a Markov random field given the marginal distribution of colors.

Many of these methods have versions in which the site to be altered is selected systematically. (In an image made up of an $M \times M$ lattice of pixels one might scan the pixels in a TV-scan.) We no longer have a stationary Markov chain. Convergence to a Markov random field as equilibrium distribution can be shown from the theory of nonhomogeneous finite Markov chains or by direct arguments.

The only evidence for preferring one of these methods to another is empirical. Folkore generally prefers systematic scans of the sites and the "Gibbs sampler".

EXERCISES

4.1. Show by direct calculation that the joint distribution of $(V_1, \ldots, V_n)$ has pdf $n!$ on $0 < v_1 < v_2 < \cdots < v_n < 1$, where $V_i = S_i/S_{n+1}$ and $S_k = E_1 + \cdots + E_k, E_i \sim \exp(1)$.

4.2. Consider various methods to simulate

    (i)   the interquartile range

    (ii)   an α-trimmed mean (the mean excluding $X_{(1)}, \ldots, X_{[n\alpha]}$ and $X_{n+1-[n\alpha]}, \ldots, X_n$)

    (iii) a trimean (a weighted average of the median and upper and lower α-percentile)

from a Cauchy distribution.

4.3. Prove that Bartlett's decomposition for the Wishart works. [Let $\mathbf{Z}$ be the $n \times p$ matrix with row $k = \mathbf{Z}_k$. Then V_{ij} is the inner product of columns i and j of $\mathbf{Z}$. These columns are independent and have a $N(0, I)$ distribution, so the distribution of W will be unchanged by rotations of the columns in $\mathbf{R}^n$. We can reduce $\mathbf{Z}$ to upper-triangular form by rotations. Then V has the required form and the diagonal elements are the lengths of $N(0, I_{n+1-i})$ vectors.]

4.4. Simulate a Poisson process on (0, 1) with rate function $\exp(\alpha + \beta t)$ by several methods and compare them [cf. Lewis and Shedler (1976)].

4.5. Use thinning to simulate a Poisson process with rate functions $\lambda_0 + \alpha \sin \omega t$, using pretesting. [$\lambda(t) \geqslant \lambda_0 - \alpha$ is obvious but better pretests are possible.]

4.6. Simulate the queueing system of Chapter 1 with one of the disciplines given there, and customer baulking at queue size 10.

4.7. Brownian motion $B(t)$ on $[0, \infty)$ has $C(s, t) = \min(s, t)$. Brownian bridge B_0 on $[0,1]$ is defined as the distribution of $B(t)$ conditional on $B(1) = 0$. Show B_0 has $C(s, t) = s(1 - t)$ for $s < t$ and that this is also true (up to a scale factor) of $B(t) - tB(1)$ on (0, 1) as well as the process $Y(\ )$ defined in Section 4.5.

CHAPTER 5

Variance Reduction

Careful design of a simulation experiment can almost always improve its effectiveness for a given cost, or reduce its cost for prescribed effectiveness. That is, the cost in *computer* time, for it is possible for the thought in the design process to outweigh the savings (as is the case for all the examples of this chapter). This suggests that we should be looking for variance reductions of at least a factor of 2 and preferably 10 or more. Another factor to bear in mind is the ubiquitous $1/\sqrt{n}$ law of statistical variation, so to reduce the standard error of an estimator by a factor of f one needs to increase the size of the experiment by around f^2. This means that large increases in computer power are needed to produce relative modest increases in precision. Another consequence is that it is conventional to quote variance reduction, not standard error reduction, as the cost reduction should be roughly proportional to the variance reduction.

How then can we achieve appreciable variance reductions? Many of the standard techniques are adaptations of ideas from sampling theory or the design of experiments. Both these subjects are of interest for simulation and can help suggest further dodges. Many techniques fall into one of the following categories.

(a) *Importance sampling.* This involves using a distribution different from the one specified in the problem, and is used to place samples where they will be most beneficial. It is closely related to *stratified* sampling.

(b) *Control and antithetic variates.* Both are devices to exploit correlations. Antithetic variates are deliberately induced to have negative correlations, which reduces the variance of their mean over that of independent samples. Control variates is regression estimation in sampling; the sample values of the variable under study are regressed on those of other variables of known mean.

(c) *Conditioning.* Exploiting the conditional structure of a problem may help us do some of the averaging analytically. We saw an example from Andrews et al. (1972) in Chapter 1.

(d) *Common variates.* This entails reusing the random-number stream during the experiment and is usually an example of the idea of "blocks" from the design of experiments.

Simulation has a colorful language, and variance reduction techniques, especially clever ones, are often known as *swindles.* Presumably it is nature that is being swindled, but she frequently gets her own back. Variance reduction swindles quite frequently do not work, especially when more than one idea is tried simultaneously. Even when variance reduction is achieved it may be at the cost of a much more complicated analysis of the resulting experiment because of the correlations induced. Variance reduction obscures the essential simplicity of simulation, which may help explain its lack of use. Despite all these caveats there is a clear responsibility on simulators to think about variance reduction when conducting a large simulation study.

5.1. MONTE-CARLO INTEGRATION

The object of any simulation study is the estimation of one or more expectations of the form $E\phi(X)$. (This may not be obvious at first sight, but even a cumulative distribution function is a collection of expectations.) Thus we can regard the problem as evaluating a frequently complex and high-dimensional integral. It is not then surprising that simulation has been used to estimate integrals with no mention of a stochastic model. This is known as Monte-Carlo integration and is one of the most fruitful fields for variance reduction techniques.

Suppose we wish to evaluate

$$\theta = \int \phi(x)f(x)dx \tag{1}$$

where x could be multidimensional (but will be scalar in all our examples). For scalar x the usual way to evaluate θ will be analytically or via numerical integration. If $x \in \mathbf{R}^2$, we can use repeated integration, but this will be awkward if $\{f > 0\} \cap \{\phi > 0\}$ is a complex shape. In higher dimensions the problems of applying numerical integration formulas become worse, and Monte-Carlo integration becomes the preferred method.

In contrast, Monte-Carlo integration is as easy in 10 dimensions as in 1. Suppose (without real loss of generality) that f is a pdf. Sample $X_1, \ldots, X_n$ independently from f and form

$$\hat{\theta} = n^{-1} \sum_1^n \phi(X_i)$$

Then $\hat{\theta}$ estimates θ, for $E\hat{\theta} = n^{-1}\sum E\phi(X_1) = E\phi(X_1) = \int \phi f\,dx$. However,

$$\text{var}(\hat{\theta}) = n^{-1}\int \{\phi(x) - \theta\}^2 f(x)dx$$

so the precision of $\hat{\theta}$ is proportional to $1/\sqrt{n}$. This contrasts markedly with numerical integration, which can use n points to achieve a precision of $O(n^{-4})$ or better. Although we cannot alter $\text{var}(\hat{\theta}) = c/n$, we can try to reduce c.

Example. Let $\theta = P(C > 2)$, where C is a Cauchy deviate. The "obvious" way to estimate θ is to let $f = 1/\pi(1 + x^2)$, the Cauchy density, and $\phi(x) = I(x > 2)$. This amounts to generating n Cauchy variates and taking $\hat{\theta}$ as the proportion greater than 2. We know

$$\theta = 1 - F(2) = \tfrac{1}{2} - \pi^{-1}\tan 2 \approx 14.76\%$$

Also $n\hat{\theta} \sim \text{binomial}(n, \theta)$ so $\text{var}(\hat{\theta}) = \theta(1 - \theta)/n \approx 0.126/n$. We will use this as our reference value for variance reductions.

It is only slightly less obvious to compute $\theta = \tfrac{1}{2}P(|C| > 2)$ by taking $\phi(x) = \tfrac{1}{2}I(|x| > 2)$. This gives $2n\hat{\theta} \sim \text{binomial}(n, 2\theta)$, so $\text{var}(\hat{\theta}) \approx 0.052/n$, a variance reduction of 2.4 times.

Another idea is to note that

$$1 - 2\theta = \int_{-2}^{2} f(x)dx = 2\int_{0}^{2} f(x)dx$$

and apply Monte-Carlo integration with $X_i \sim U(0,2)$ and $\phi(x) = 2f(x)$. This gives $\text{var}(\hat{\theta}) \approx 0.028/n$. Alternatively, note that for $y = 1/x$,

$$\theta = \int_{2}^{\infty} \frac{1}{\pi(1 + x^2)}dx = \int_{0}^{1/2} \frac{y^{-2}dy}{\pi(1 + y^{-2})} = \int_{0}^{1/2} f(y)dy$$

a coincidence! Using $X_i \sim U(0, \tfrac{1}{2})$, $\phi(x) = f(x)/2$ gives $\text{var}(\hat{\theta}\to) \approx 9.3 \times 10^{-5}/n$, a variance reduction of 1350.

This gives us four forms of the problem without using any of the general techniques. Table 5.1 illustrates the variance reductions obtainable. The largest value corresponds to $\text{var}(\hat{\theta}) \approx 1.1 \times 10^{-9}/n$ and removes the need for the experiment. This example is of course rather unrealistic as the answer is known or could have been found easily by numerical integration. $\square$

Table 5.1. Variance Reductions Obtainable for $\theta = P(C > 2)$

	θ			
	$P(C > 2)$	$P(|C| > 2)$	$1 - \int_0^2 f$	$\int_0^{1/2} f$
Straight	1	2.4	20	1350
Hit-or-miss	—	—	1.6	—
Importance	1350	—	—	—
Control variates	—	—	7400	1.1×10^8
Antithetic variates	2.4	18	24	12

There is an even less efficient version of Monte-Carlo integration known as "hit-or-miss Monte-Carlo." Consider a bounded function ϕ on (a, b), with $0 \leqslant \phi \leqslant c$. Then θ is the area under the curve, and can be estimated as $c(b - a)$ times the proportion under ϕ of $[a, b] \times [0, c]$. Let $U \sim U(a, b)$ and $V \sim U(0, c)$. Then

$$\theta = c(b - a)P(V \leqslant \phi(U))$$

Take n independent samples (U_i, V_i) and let $\tilde{\theta}$ be $c(b - a)$ times the proportion with $V_i \leqslant \phi(U_i)$. Then $E\tilde{\theta} = \theta$ and $\text{var}(\tilde{\theta}) = \theta[c(b - a) - \theta]/n$. Table 5.1 shows that in our standard example $\text{var}(\tilde{\theta}) > \text{var}(\hat{\theta})$, and this is true generally.

Theorem 5.1. Suppose $\theta = \int_a^b \phi(x)dx$ with $0 \leqslant \phi \leqslant c$. Then $\text{var}(\hat{\theta}) \leqslant \text{var}(\tilde{\theta})$ with equality only if $\phi \equiv c$, when $\text{var}(\hat{\theta}) = 0$.

PROOF.

$$n \, \text{var}(\hat{\theta}) = \int_a^b \phi(x)^2(b - a)dx - \theta^2 \text{ taking } f \text{ uniform on } (a, b)$$

$$\leqslant c\int_a^b \phi(x)(b - a)dx - \theta^2$$

$$= c(b - a)\theta - \theta^2 = n \, \text{var}(\tilde{\theta})$$

with equality only if $\phi \equiv c$.   □

This shows hit-or-miss should never be used, although it sometimes is.

5.2. IMPORTANCE SAMPLING

The idea of importance sampling is related to weighted and stratified sampling ideas in sampling theory. Suppose we wish to estimate $\theta = E\phi(X)$ for an observation X on a random system. Then some outcomes of X may be more important that others in determining θ, and we would wish to select such values more frequently. A simple example is to take θ as the probability of the occurrence of a very rare event. Then the only way to estimate θ at all accurately may be to produce the rare events more frequently.

Suppose we simulate a model which gives pdf g to X rather than the correct pdf f, and both pdfs are known. Then if we let $\psi = \phi f/g$,

$$\theta_g = n^{-1}\sum \psi(X_i), \qquad X_i \sim g$$

is an unbiased estimator of θ. This is a weighted mean of the $\phi(X_i)$ with weights inversely proportional to the "selection factor" g/f. We have

$$\mathrm{var}(\theta_g) = n^{-1}\int \{\psi(x) - \theta\}^2 g(x)dx = n^{-1}\int \left(\frac{\phi f}{g} - \theta\right)^2 g\,dx$$

which can be much smaller than $\mathrm{var}(\theta)$ provided g is chosen to make $\phi f/g$ nearly constant. It is easy to show that $\mathrm{var}(\theta_g)$ is minimized by $g \propto |\phi f|$ (Exercise 5.3), but this is impracticable. It does however give us an idea of how to choose g.

Consider our standard example, $\theta = P(C > 2)$. We select g so that $\{g > 0\} = \{|\phi f| > 0\} = \{x > 2\}$. On $x > 2$, $f(x)$ is closely matched by $g(x) = 2/x^2$. We can easily sample from g by inversion to get $X = 2/U$. Then

$$\psi(x) = 2f(x)/g(x) = 2x^2/\pi(1 + x^2) = 2/\pi(1 + x^{-2})$$

and $X^{-1} \sim U(0, \tfrac{1}{2})$. Thus importance sampling here reduces to Monte-Carlo integration of $\int_0^{1/2} f(y)dy$.

Example (Siegmund, 1976). Let $X_1, X_2, \ldots$ be independent identically distributed random variables with partial sums $S_n = X_1 + \cdots + X_n$. For $a \leqslant 0 < b$ let

$$T = \min\{n \,|\, S_n \leqslant a \text{ or } S_n \geqslant b\}$$

The theory of sequential tests needs probabilities such as $\theta = P(S_T \geqslant b)$. Siegmund showed how importance sampling could be used to estimate θ more precisely.

Table 5.2. Results of a Simulation Experiment to Find $\theta = P(S_T \geqslant b)$ for $T = \min\{n | S_n \leqslant a \text{ or } S_n \geqslant b\}$ with $S_n = X_1 + \cdots + X_n$, $X_i \sim N(\mu, 1)$ and $a = -4, b = 7$

μ	Direct θ	s.e.	$\tilde{\theta}$	s.e.	Variance Reduction
0	0.389	0.0049	0.389	0.0055	1
−0.1	0.149	0.0035	0.147	0.0010	12
−0.2	0.041	0.0020	0:0412	1.8×10^{-4}	110
−0.3	0.011	0.0010	0.00996	3.8×10^{-5}	750
−0.5	0.0005	0.0007	0.000505	2.3×10^{-6}	9,600

For simplicity, consider $X_i \sim N(\mu, 1)$. By symmetry we need only consider $\mu \leqslant 0$. [If $\mu > 0$, estimate $\theta = 1 - P(S_T \leqslant a)$.] Importance sampling should increase the chance that $S_T \geqslant b$, which we ought to achieve by taking a new mean $v > 0$. Choosing $v = -\mu$ gives both algebraic simplicity and a certain optimality (loc.cit.) Then the ratio of the pdfs is $\exp(2\mu S_T)$, so the estimator of θ under importance sampling is

$$\tilde{\theta} = n^{-1} \sum I(Y_i \geqslant b)\exp(2\mu Y_i)$$

where Y_i are independent realizations of S_T. The summand is less than $\exp(2\mu b)$ (remember $\mu \leqslant 0$), which helps explain its reduced variability. Table 5.2 shows the results of a small experiment with $a = -4$, $b = 7$, and $n = 10^4$. Using importance sampling made negligible difference to the total cost.

Siegmund gives a further idea for $a = -b$. Then for $X_i \sim N(-\mu, 1)$ both $I(S_T \geqslant b) \exp(2\mu S_T)$ and $I(S_T \leqslant -b)$ have mean θ and they should be negatively correlated. This suggests

$$\theta^* = (1 - \lambda)I(S_T \geqslant b)\exp(2\mu S_T) + \lambda I(S_T \leqslant -b)$$

with λ chosen to approximately minimize $\text{var}(\theta^*)$. This is similar to the ideas of Section 5.3. □

5.3. CONTROL AND ANTITHETIC VARIATES

Both control and antithetic variates aim to reduce the variability of an estimator by using two quantities that vary together. Suppose we wish to estimate $\theta = EZ$ for some $Z = \phi(X)$ observed on a process X. A *control variate* is another observation $W = \psi(X)$, which we believe varies with Z and which has a known mean. We can then estimate θ by averaging observations of $Z - (W - EW)$. A simple example is to estimate the mean of a sample median using the sample mean as control variate. *Antithetic variates* come in pairs. Suppose Z^* has the same distribution as Z but is negatively correlated with Z. Suppose we estimate θ by $\tilde{\theta} = \frac{1}{2}(Z + Z^*)$. Clearly $\tilde{\theta}$ is unbiased, and

$$\text{var}(\tilde{\theta}) = [2 \, \text{var}(Z) - 2 \, \text{cov}(Z, Z^*)]/4$$
$$= \frac{1}{2} \, \text{var}(Z) \, [1 + \text{corr}(Z, Z^*)]$$

Thus we will obtain a more precise estimator from n pairs (Z, Z^*) than $2n$ observations of Z provided $\text{corr}(Z, Z^*) < 0$. It might be cheaper to observe (Z, Z^*) than observe Z twice, in which case we gain even more.

Example (Rothery, 1982).   Suppose we wish to compare the power functions of two hypothesis tests S and T. (Here S and T denote the indicator functions of the critical regions, so power $= ES$.) If the power function of T can be found analytically, we can use T as a control variate. If S and T are comparable in performance, we will frequently have $S = T = 0$ or $S = T = 1$. Rothery found variance reduction factors of 2–6 in his example.   □

The theory as presented above is straightforward. The problem arises in identifying suitable variates. Generally we will have more than one candidate for a control variate, and only know that Z generally varies with each W_i. This suggests

$$\theta = Z - \beta_1(W_1 - EW_1) - \cdots - \beta_p(W_p - EW_p)$$

as an unbiased estimator of θ. The coefficients β are chosen to minimize $\text{var}(\theta)$. Consider first the case $p = 1$. Then

$$\text{var}(\theta) = \text{var}(Z) - 2\beta \, \text{cov}(Z, W) + \beta^2 \, \text{var}(W)$$

so we should take $\beta = \text{cov}(Z, W)/\text{var}(W)$. It is unlikely that we would know $\text{cov}(Z, W)$ and not $E(Z)$, and we might or might not know $\text{var}(W)$. Replacing $\text{cov}(Z, W)$ and $\text{var}(W)$ by their sample equivalents amounts to regressing the observations of Z on the observations of W. The general case is similar and leads to multiple regression of Z on $W_1, \ldots, W_p$.

This is not a standard regression problem because we have random regressors W_i. Thus we do not necessarily obtain an unbiased estimator of θ or var($\hat\theta$) by standard regression methods. However, if the coefficients $\hat\beta_i$ are found by a preliminary experiment and are held fixed, we will obtain unbiased estimators, for if $W_i' = W_i - EW_i$,

$$E\hat\theta = n^{-1}[EZ - \sum \beta_j E(W_j')] = \theta$$

$$E\sum(Z_i - \sum \beta_i W_{ij}')^2 = nE(Z - \sum \hat\beta_j W_j')^2$$

$$\text{var}(\hat\theta) = n^{-1}E(Z - \sum \hat\beta_j W_j')^2$$

so $n^{-2} RSS$ is an unbiased estimator of var($\hat\theta$).

Example.   We return to our standard example in the form $\theta = \frac{1}{2} - \int_0^2 f(x)dx$. Expanding $f(x) = 1/\pi(1 + x^2)$ suggests control variates x^2 and x^4. A small regression experiment gave

$$\hat\theta = \frac{1}{2} - [f(X) + 0.15(X^2 - 8/3) - 0.025(X^4 - 32/5)]$$

and var($\hat\theta$) $\approx 6.3 \times 10^{-4}/n$. Plotting $f(x)$ and this quartic suggests that we would do better by including terms in x and x^3. Doing so reduces var($\hat\theta$) to $3.8 \times 10^{-6}/n$, a variance reduction factor of 7400.

We can also apply control variates to $\theta = \int_0^{1/2} f(x)dx$. Fitting x^2 and x^4 we find

$$\hat\theta = f(X) + 0.312(X^2 - 1/24) - 0.233(X^4 - 1/160)$$

which gives var($\hat\theta$) $\approx 1.1 \times 10^{-9}/n$. In other words, $\hat\theta$ is effectively constant, and one sample will give an accurate enough approximation to its value.   □

We can explore further the properties of control variates by assuming joint normality of $(Z, W_1, \ldots, W_p)$. Many control variates are sums or averages, so this may not be an unreasonable assumption. We will also assume that $EW_i = 0$. How much do we lose by estimating β in a prior experiment? By joint normality there is an independent variate U and coefficients β_i such that

$$Z = \theta + U + \sum \beta_i W_i = \theta + U + \mathbf{w}\beta$$

where $\mathbf{w} = (W_1, \ldots, W_p)$. Clearly $\mathbf{w}\beta$ is the optimal control leading to

$$\text{var}(\hat\theta) = \sigma_U^2/n = \text{var}(Z|\mathbf{w})/n$$

If β is estimated by $\hat{\beta}$

$$\mathrm{var}(\hat{\theta}|\hat{\beta}) = n^{-1}\, \mathrm{var}(Z - \mathbf{w}\hat{\beta})$$

$$= n^{-1}\, \mathrm{var}(U - \mathbf{w}(\hat{\beta} - \beta))$$

$$= n^{-1}[\sigma_U^2 + \mathrm{var}(\mathbf{w}(\hat{\beta} - \beta))]$$

by the independence of U and $\mathbf{w}$. Clearly

$$E(\hat{\theta}|\hat{\beta}) = E[Z - \mathbf{w}\hat{\beta}] = \theta - 0\hat{\beta} = \theta$$

Consider $p = 1$. Then

$$\mathrm{var}(\hat{\theta}) = \sigma_U^2/n + n^{-1}\sigma_W^2\, \mathrm{var}(\hat{\beta})$$

Suppose β was estimated from t runs. Then

$$\hat{\beta} = \sum(w_i - \bar{w})Z_i / \sum(w_i - \bar{w})^2$$

$$\mathrm{var}(\hat{\beta}|\{w_i\}) = \sigma_U^2 / \sum(w_i - \bar{w})^2$$

$$\mathrm{var}(\hat{\beta}) = \sigma_U^2 E[1/\sum(w_i - \bar{w})^2]$$

and $\sum(w_i - \bar{w})^2 \sim \sigma_W^2 \chi_{t-1}^2$, so

$$\mathrm{var}(\hat{\beta}) = \sigma_U^2/\sigma_W^2(t - 3)$$

$$\mathrm{var}(\hat{\theta}) = \sigma_U^2 n^{-1}[1 + (t - 3)^{-1}]$$

Now suppose $p > 1$, and let $\mathbf{W}$ be the $t \times p$ matrix of observations of $\mathbf{w} - \bar{\mathbf{w}}$ in the preliminary experiment. We find

$$\hat{\beta} = (\mathbf{W}^T\mathbf{W})^{-1}\mathbf{W}^T\mathbf{Z}$$

$$\mathrm{var}(\hat{\beta}|\mathbf{W}) = \sigma_U^2(\mathbf{W}^T\mathbf{W})^{-1}$$

$$\mathrm{var}(\hat{\beta}) = \sigma_U^2 \Sigma^{-1}/(t - p - 2)$$

where Σ is the covariance matrix of $\mathbf{w}$ (using properties of the Wishart distribution). Thus

$$\mathrm{var}(\hat{\theta}) = \sigma_U^2 n^{-1}[1 + E(\mathbf{w}\Sigma^{-1}\mathbf{w}^T)/(t - p - 2)]$$

$$= \sigma_U^2 n^{-1}[1 + p/(t - p - 2)]$$

$$= \frac{\sigma_U^2(t - 2)}{n(t - p - 2)}$$

The loss is thus negligible provided $t \gg p$.

Now suppose $\hat{\beta}$ is estimated from the n observations of the actual experiment. That is, we fit

$$Z_i = \theta + w_i \hat{\beta}$$

from n observations by a regression. Then

$$\theta = n^{-1} \sum [Z_i - w_i \hat{\beta}] = n^{-1} \sum [U_i - w_i(\hat{\beta} - \beta)] + \theta$$

so

$$E[\theta | \{w_i\}] = \theta$$

and

$$\text{var}(\theta | \{w_i\}) = \sigma_U^2 (X^T X)_{11}^{-1}$$

where the rows of X are $(1, w_i)$. Finally,

$$\theta \sim N(\theta, \sigma_U^2 (X^T X)_{11}^{-1}) \quad \text{conditional on } \{w_i\}$$

so we can find a conditional $(1 - \alpha)$-confidence interval by

$$(\theta - t_\alpha c \hat{\sigma}_U, \; \theta + t_\alpha c \hat{\sigma}_U)$$

where $c = \sqrt{(X^T X)_{11}^{-1}}$, $\hat{\sigma}_U^2 = RSS/(n - p - 1)$ and t_α is the upper $(1 - \frac{1}{2}\alpha)$ point of t_{n-p-1}. Since this is a valid confidence interval conditionally, it is also a valid unconditional $(1 - \alpha)$-confidence interval. [This argument follows Cheng (1978b) and Lavenberg et al. (1982).]

How much do we lose by estimating β? We have

$$\text{var}(\theta | \{w_i\}) = \sigma_U^2 (X^T X)_{11}^{-1} = \sigma_U^2 \left[\frac{1}{n} + \bar{w}(W^T W)^{-1} \bar{w}^T \right]$$

$$> \sigma_U^2/n$$

so we *do* lose. Consider first $p = 1$. Then

$$\text{var}(\theta | \{w_i\}) = \sigma_U^2 n^{-1} [1 + n\bar{w}^{-2}/\sum (w_i - \bar{w})^2]$$

Now $\sum(w_i - \bar{w})^2 \sim \sigma_W^2 \chi_{n-1}^2$ independently of $\bar{w}$, so

$$\begin{aligned}
\mathrm{var}(\hat{\theta}\Omega &= \sigma_U^2 n^{-1}[1 + \sigma_W^2/\sigma_W^2(n-3)] \\
&= \sigma_U^2 n^{-1}[1 + 1/(n-3)] \\
&= \sigma_U^2 \frac{(n-2)}{n(n-3)}
\end{aligned}$$

For $p > 1$ we find

$$\mathrm{var}(\hat{\theta}|\{w_i\}) = \sigma_U^2\left[\frac{1}{n} + \bar{w}(W^T W)^{-1}\bar{w}^T\right]$$

$$\mathrm{var}(\hat{\theta}|\bar{w}) = \sigma_U^2 n^{-1}[1 + n\bar{w}\Sigma^{-1}\bar{w}^T/(n-p-2)]$$

$$\mathrm{var}(\hat{\theta}) = \sigma_U^2 n^{-1}[1 + p/(n-p-2)]$$

$$= \frac{\sigma_U^2}{n}\frac{n-2}{n-p-2}$$

using the independence of $\bar{w}$ and $(W^T W)$ and properties of the Wishart distribution. Thus the loss in estimating β is the factor $(n-2)/(n-p-2)$ and is negligible if $n \gg p$.

These calculations are all done assuming joint normality. Without this $\hat{\theta}$ may be biased, and the use of jackknife and bootstrap techniques (Section 7.1) has been suggested to reduce the bias. Lavenberg et al. (1982) give some examples, but prefer the confidence interval based on joint normality.

When Σ is known we could consider using it in the estimation of θ. Consider estimating θ in

$$Z = \theta + w\beta$$

by maximum likelihood under joint normality. Then the log likelihood is

$$\text{const} - \frac{1}{2\sigma_U^2}\sum(Z - \theta - w\beta)_i^2 - \tfrac{1}{2}\sum(w_i\Sigma^{-1}w_i^T)$$

and the maximum likelihood estimators of θ and β are those given above when Σ is unknown. If we insisted on replacing the sample variance of w by Σ we would obtain a conditionally biased $\hat{\beta}$ and much more complicated inferences. Cheng (1978b) and Cheng and Feast (1980b) point out that it *is* advantageous to use Σ if var(Z) is to be estimated, but this is of secondary interest (if any).

Antithetic Variates

The method of antithetic variates finds two correlated estimators of θ and combines them. Suppose X and X* are two outcomes of our model, giving rise to $Z = \phi(X)$ and $Z^* = \phi(X^*)$. Then

$$\text{var}[\tfrac{1}{2}(Z + Z^*)] = \tfrac{1}{2}\, \text{var}(Z)[1 + \text{corr}(Z, Z^*)]$$

so we obtain a smaller variance on averaging Z and Z^* rather than two independent realizations provided $\text{corr}(Z, Z^*) < 0$, and a substantial reduction if we can achieve large negative correlation.

A standard way to achieve this correlation in simple models is to generate Z and Z^* by inversion as $Z = F^-(U), Z^* = F^-(1 - U)$. Then $\text{corr}(Z, Z^*) < 0$ by Theorem 5.2.

Theorem 5.2. Suppose g is a monotonic function on $(0, 1)$. Then

$$\text{corr}(g(U), g(1 - U)) < 0$$

PROOF. Without loss of generality assume g is increasing. Let $\theta = Eg(U)$, and $t = 1 - \inf\{u | g(u) > \theta\}$. Then

$$\text{cov}(g(U), g(1 - U)) = Eg(U)[g(1 - U) - \theta]$$

$$= \int_0^t g(u)[\underset{>0}{g(1 - u)} - \theta]du$$

$$+ \int_t^1 g(u)[\underset{>0}{g(1 - u)} - \theta]du$$

$$< g(t) \int_0^1 g(1 - u) - \theta\, du = 0 \qquad \square$$

For symmetric distributions we can obtain perfect negative correlation [but $EZ = F^-(0.5) =$ point of symmetry is obvious anyway]. Consider the Bernoulli distribution with $P(Z = 1) = 1 - P(Z = 0) = p$. Then

$$\max\left[\frac{-p}{(1 - p)}, \frac{-(1 - p)}{p}\right] \leqslant \text{corr}(Z, Z^*) \leqslant 1$$

for any random variable Z^* with the same distribution (Exercise 5.4), so for p near zero or one only minor variance reduction is possible. Another well-known example is $Z \sim \exp(\lambda)$ for which $\text{corr}(Z, Z^*) \approx -0.645$ (Page, 1965).

The following result (Hoeffding, 1940; Fréchet, 1951; Whitt, 1976) shows that we *do* achieve the best possible negative correlation.

Theorem 5.3. Suppose X and Y have a common marginal distribution with cdf F. Then

$$\max(0,\ F(x) + F(y) - 1) \leqslant P(X \leqslant x, Y \leqslant y) \leqslant F(\min(x, y))$$

with both extremes being attained and giving minimal and maximal values of $\text{corr}(X, Y)$. The upper bound corresponds to $Y = X$, the lower to $X = F^{-1}(U)$, $Y = F^{-}(1 - U)$, $U \sim U(0, 1)$.

PROOF. $P(X \leqslant x, Y \leqslant y) \leqslant P(X \leqslant x) \leqslant F(x)$. By symmetry $P(X \leqslant x, Y \leqslant y) \leqslant \min(F(x), F(y)) = F(\min(x, y))$, which is clearly attained by $X = Y$. For the lower bound

$$P(X \leqslant x, Y \leqslant y) = P(X \leqslant x) - P(X \leqslant x, Y > y)$$
$$\geqslant F(x) - P(Y > y) = F(x) + F(y) - 1$$

To see that the bound is attained, consider

$$P(F^{-}(U) \leqslant x, F^{-}(1 - U) \leqslant y) = P(U \leqslant F(x), 1 - U \leqslant F(y))$$
$$= P(1 - F(y) \leqslant U \leqslant F(x)) = \max(0, F(x) + F(y) - 1)$$

Let $H(x, y) = P(X \leqslant x, Y \leqslant y)$. We will show

$$\text{cov}(X, Y) = \iint [H(x, y) - F(x)F(y)]dx\,dy$$

whence the extreme joint cdfs give extreme correlations. Consider independent pairs (X_1, Y_1), (X_2, Y_2) with joint cdf H. Then

$$E[(X_1 - X_2)(Y_1 - Y_2)] = E[X_1 Y_1 - X_1 Y_2 - X_2 Y_1 + X_2 Y_2]$$
$$= 2EXY - 2EXEY = 2\,\text{cov}(X, Y)$$

The left-hand side is

$$E \iint [I(u \leqslant X_1) - I(u \leqslant X_2)][I(v \leqslant Y_1) - I(v \leqslant Y_2)]du\,dv$$
$$= \iint E[I(u \leqslant X_1) - I(u \leqslant X_2)][I(v \leqslant Y_1) - I(v \leqslant Y_2)]du\,dv$$

$$= \iint [H(u, v) - F(u)F(v) - F(u)F(v) + H(u, v)] \, du \, dv$$

$$= 2 \iint [H(u, v) - F(u)F(v)] \, du \, dv$$

as required. $\square$

This theorem rather restricts the applicability of antithetic variates. For Monte-Carlo integration on (a, b), $\theta = \int \phi \, dx$, we find

$$\tilde{\theta} = \sum \{\phi(a + (b - a)U_i) + \phi(b - (b - a)U_i)\}(b - a)/2n$$

with

$$\text{var } \tilde{\theta} = \frac{b - a}{4n} \int_0^1 \{\psi(u) + \psi(1 - u)\}^2 \, du - \theta^2/n$$

versus

$$\text{var } \tilde{\theta} = n^{-1}(b - a) \int_0^1 \psi(u)^2 \, du - \theta^2/n$$

where $\psi(u) = \phi(a + (b - a)u)$. We see that var $\tilde{\theta}$ will be small if $\{\psi(u) + \psi(1 - u)\}$ is nearly constant, in particular if ψ and hence ϕ is nearly linear.

Example. Direct application of antithetic variates to our standard example merely changes C to $-C$ and so achieves the same variance reduction as counting $|C| > 2$. Suppose we generate samples of $|C|$ by $\tan(\pi U/2)$. Here antithetic variates again counts half the number of events of double probability and so has var$(\tilde{\theta}) \approx 0.030/n$. For the alternative form $\theta = \frac{1}{2} - \int_0^2 f \, dx$ we find var$(\tilde{\theta}) \approx 5.9 \times 10^{-4}/n$, although this requires $2n$ evaluations of f and should be compared with earlier formulas evaluated at $2n$. For $\theta = \int_0^{1/2} f \, dx$ we find var$(\tilde{\theta}) \approx 3.8 \times 10^{-6}/n$. These last two cases give variance reductions of 24 and 12, reflecting the greater linearity of f over $(0, 2)$ compared with $(0, \frac{1}{2})$. $\square$

For Monte-Carlo integration of a smooth function ϕ we can always obtain approximate linearity by splitting the range of integration into small parts. Thus we can consider using first stratified sampling to divide up the range of integration, then antithetic variates on the pieces. This is bound to lead to large variance reductions at the expense of additional programming.

It is, however, beginning to look like a simple numerical integration routine (the trapezoidal rule) and will still be outclassed by numerical integration.

Antithetic variates are more difficult to apply when less is known about the problem than in unidimensional Monte-Carlo integration. There are two problems in its use. First, the method suggested by Theorem 5.2 to induce negative correlations is inversion. Inversion may be tedious or impracticable, yet it is very much more difficult to achieve appreciable negative correlation with other methods of generation. Second, in more complex models such as queueing systems we do not generate directly the variables of interest (such as waiting times and queue lengths).

Cheng (1982) takes a novel approach to the use of antithetic variates in queueing systems. He takes the point of view that what are negatively correlated in most attempts at antithetic variates are control variates, and inducing negative correlation between the control variates can induce positive correlation between their prediction errors. That is, if

$$Z = \theta + C + \eta$$

then achieving corr$(C, C^*) < 0$ may cause corr$(\eta, \eta^*) > 0$. Cheng's approach is to construct a simulation that differs as little as possible from X while replacing C by C^*. It is best seen in a simple example. Suppose we have a queue with exponential service times and are interested in mean waiting time. We would expect the waiting time to increase with the average service time. Let $S = \bar{S_i}$ denote the average of the service times. For the antithetic simulation we construct new service times σ_i and rescale them to have mean $S^* = F^-(1 - F(S))$, where F is the cdf of S. The effect of this should be to induce negative correlation in the mean waiting times of the two simulations. The reader is referred to the original paper for further details and extensions. Note that in almost all cases it is an approximate technique in that the two simulations do not have exactly the same distribution.

Queueing Systems

There has been a fair amount of attention to the use of control and antithetic variates in queueing systems. Two obvious choices for control variates are the mean arrival and mean service times. Other possibilities are considered by Carson and Law (1980), Iglehart and Lewis (1979), Kleijnen (1974/5), Lavenberg and Welch (1981), and Lavenberg et al. (1982).

For antithetic variates we can apply the $U \rightarrow 1 - U$ transformation to the arrival times, service times or both. Page (1965) suggested interchanging the random numbers used for arrival and service times. Mitchell (1973) used an extended version of Theorem 5.2 to show that when estimating steady-state waiting time in a GI/G/1 FIFO queue both interchanging random numbers and applying antithetic variates to both arrival and service times simultaneously reduce the estimation variance. Such results are rare, and often a pilot experiment is needed to see if an idea for control or antithetic variates is beneficial.

Example. Consider the GI/G/1 queue with FIFO service. This has a general cdf G of service times S_i and a renewal process of arrivals with interval times A_i of cdf F. There is one server who serves customers in order of arrival. Let W_i denote the wait (excluding service time) experienced by the ith customer to arrive. Suppose customer 1 arrives at time 0 at an empty queue. Then

$$W_1 = 0$$
$$W_i = \max(0, W_{i-1} - A_i + S_{i-1}), \quad i \geqslant 2$$

which recursively determines the waiting times from (A_i, S_i). Note that (W_i) are not independent.

Suppose we wish to estimate $\omega = EW_{100}$. One suggestion for a control variate is defined by

$$C_1 = 0$$
$$C_i = C_{i-1} - A_i + S_{i-1}, \quad i \geqslant 2$$

so $C_i = $ (sum of service times of past customers − arrival time of ith customer) and has a known mean. If the queue is usually busy we can expect $W_i \approx C_i$.

Table 5.3. Estimated Standard Errors of Estimators of $\omega = EW_{100}$ in a M/M/1 Queue with Traffic Intensity 0.9[a]

	Standard Error	Variance Reduction
Straight	$5.8/\sqrt{n}$	—
Control variate C_{100}	$4.0/\sqrt{n}$	2.1
Antithetic variates		
$\quad U_i \leftrightarrow V_i$	$3.0/\sqrt{n}$	3.6
$\quad U_i \rightarrow 1 - U_i$	$5.9/\sqrt{n}$	<1
$\quad V_i \rightarrow 1 - V_i$	$6.0/\sqrt{n}$	<1
$\quad$ Both	$3.1/\sqrt{n}$	3.5

[a]n denotes the total number of runs.

We can generate $A_i = F^-(U_i), S_j = G^-(V_j)$. Possible antithetic schemes are shown in Table 5.3 together with their effect. In this example $A_i \sim \exp(0.9)$, $S_i \sim \exp(1)$, so $\omega \approx 6$ and $P(W_{100} = 0) \approx 12\%$. In this example most of the computing cost was in generating the exponential deviates, so the variance reductions were effectively free and would be worthwhile.

5.4. CONDITIONING

The application of conditioning depends very much on the problem under study; there is no general theory. We can always say

$$\text{var}(E[Z|W]) = \text{var}(Z) - E(\text{var}[Z|W]) \leqslant \text{var}(Z)$$

so forming *any* conditional expectation $E[Z|W]$ analytically will reduce variability. The problem, of course, is to identify the right W. The example given in Chapter 1 is of this type. The work of Andrews et al. (1972) was anticipated by Dixon and Tukey (1968) and Relles (1970), and is expounded in detail by Gross (1973) and Simon (1976). Almost all these references are restricted to estimators of location with specific distributions, although Simon also considers estimators of spread.

Burt and Garman (1971) and Garman (1972) provide an interesting example from PERT analysis. Consider a network such as Fig. 5.1. The times of transfer are random and independent; the quantity of interest is the minimum passage time T from A to B by any allowed route. Burt and Garman note that *conditional* on $T_1 = t_1$ and $T_5 = t_5$ there are three possible passage times

$$t_1 + T_4$$
$$t_1 + T_3 + t_5$$
$$T_2 + t_5$$

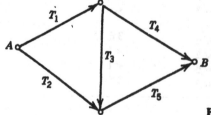

Figure 5.1. A simple stochastic PERT network.

which are independent. Thus $T = \min(t_1 + T_4, t_1 + T_3 + t_5, T_2 + t_5)$ has a known conditional cdf

$$F_X(x) = 1 - [1 - F_4(x - t_1)][1 - F_3(x - t_1 - t_5)][1 - F_2(x - t_5)]$$

where F_i is the cdf of T_i. Thus conditioning reduces the problem to a simulation experiment on (T_1, T_5).

Garman (1972) pointed out that conditional on $T_1 = t$ we have the equivalent network of Fig. 5.2, and this has a conditional cdf of X which can be computed analytically by series–parallel reduction. Let $T_6 = \min(T_2, T_3 + t)$ whose cdf can be found. Then $X = \min(T_4 + t, T_6 + T_5)$. This conditioning reduces the problem to a simulation experiment on T_1.

The network of Fig. 5.1 is rather simple, but the principle is applicable quite generally to networks.

We can use conditioning in estimating EW_{100} in the GI/G/1 queue. Chapter 1 introduced the concept of a *tour* between times at which the queue is totally empty. Each run will provide us with a random number of tours plus one observation on N, the number of the last customer before the 100th that arrived at empty queue. Then

$$\omega = EW_{100} = E[E(W_{100-N}|N)] = \sum P(N = r)w_r$$

where $w_r = E(W_i|W_i > 0, i = 2, \ldots, r - 1)$. Thus we can estimate w_r from the tours, and $P(N = n)$ from the runs. The combined estimate will make more use of the observations than merely recording W_{100}, and should be more accurate, particularly at lower traffic intensities. However, it appears to be very difficult to assess the variability of an estimator that is not an average, except by repeating the whole experiment.

Carter and Ignall's (1975) "virtual measures" are conditioning repackaged for rare events. Their motivating example was the provision of fire-fighting appliances. The rare events are the inability to supply enough machines of the right type. One way to overcome this is to simulate the typical behavior of

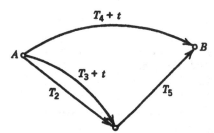

Figure 5.2. An equivalent network to Fig. 5.1 conditional on $T_1 = t$.

fighting small fires, and compute analytically the probability of a failure to cover a large fire given the state of the system, averaging this over the simulation of the system. This is a case in which var$[Z|W]$ is large, so a large variance reduction is attainable by conditioning.

The term "conditional Monte Carlo" was coined by Trotter and Tukey (1957) for a somewhat different idea. Suppose we have a space of outcomes $\mathscr{Z}$ which can be described by the pair (x, y). Conditional Monte Carlo is a way to evaluate the conditional expectation $E[\phi(X)|Y = y_0]$ without restricting sampling to the potentially awkward set $\{z(x, y_0)\}$. Let Z be a random variable taking values in $\mathscr{Z}$, and let X and Y correspond to Z, so $Z = z(X, Y)$. Let h be the pdf of (X, Y). Then we estimate $E[\phi(X)|Y = y_0]$ by $\phi(X)w(X, Y)$, where

$$w(X, Y) = \frac{h(X, y_0)}{h(X, Y)} \frac{\zeta(X, Y)}{\xi(X)f_Y(y_0)}$$

and $\zeta(x, y)$ is an arbitrary function with $\xi(x) = \int \zeta(x, y)dy \neq 0$. Then

$$E[\phi(X)w(X, Y)] = \int \phi(x)w(x, y)h(x, y)dxdy$$

$$= \int \left\{ \frac{\phi(x)h(x, y_0)}{f_Y(y_0)\xi(x)} \int \zeta(x, y)dy \right\} dx$$

$$= \int \phi(x)h(x, y_0)/f_Y(y_0)dx$$

$$= \int \phi(x)f_X(x|Y = y_0)dx$$

$$= E[\phi(X)|Y = y_0]$$

so the estimator is unbiased. Usually this is written using Z, in which case

$$w(Z) = \frac{J(Z)h'(Z_0)\zeta(Z)}{J(Z_0)h'(Z)\xi(X)f_Y(y_0)}, \qquad Z_0 = (X, y_0)$$

and $h'(z) = Jh(x, y)$ is the pdf of Z, J the Jacobian of $z \rightarrow (x, y)$. The function ζ is available for variance reduction. For fixed x we can use ζ to reduce the fluctuations in $\zeta(x, y)/h(x, y)$, so we should take ζ similar to $f_Y(y|X = x)$. Taking $\zeta(x, y) = f_Y(y|X = x)$ makes $w(x, y) = f_X(x|Y = y_0)/f_X(x)$, which reduces to importance sampling using the simpler unconditional distribution. The general method avoids calculating f_X.

Example. Suppose $X_1, \ldots, X_n$ is a $N(0, 1)$ sample, and $X_{(1)}, \ldots, X_{(n)}$ are its order statistics. Let

$$R = X_{(n)} - X_{(1)}$$
$$S = \max(X_{(n)} - X_{(2)}, X_{(n-1)} - X_{(1)})$$

In studying tests of outliers we might want to know

$$\theta = P(S \geqslant s | R = r)$$

We use the fact that R is a scale factor to define $Y = R$, $X = (X_1, \ldots, X_n)/R$ (which has only $n - 1$ degrees of freedom). Now

$$h(z) = h(x_1, \ldots, x_n) = (2\pi)^{-n/2} \exp(-\tfrac{1}{2} \sum x_i^2)$$

and $f_R(r)$ is known. Computing the Jacobian gives

$$w(Z) = \exp\{\tfrac{1}{2}(1 - \lambda^2)\|Z\|^2\} \times \lambda^{n-1} \zeta(Z)/\zeta(X) f_R(r)$$

where $\lambda = r/R$. We take

$$\phi(X) = I(S/R \geqslant s/r)$$

Then ζ is chosen to flatten w as a function of R. Since $\lambda \propto R^{-1}$ and $\|Z\|^2 \propto R^2$, this suggests $\zeta(Z) \propto R^{n-1} \exp(-cR^2)$ for some constant c chosen by experiment. $\square$

This example is a simplified version of the original application of Arnold et al. (1956).

5.5. EXPERIMENTAL DESIGN

Many simulation experiments are done to compare conditions, for example to compare queueing disciplines in a model of a bank (such as that described in Chapter 1). In such cases we can use all the ideas of the design of experiments to produce better comparisons between the different conditions. We aim only to provide an overview of what is possible, referring the reader to the literature for further details. Box et al. (1978) provide an especially convincing introduction to the subject.

One of the fundamental ideas in experimental design is the grouping of experiment units into *blocks* which are more homogeneous than the total pool of experimental units. In a simulation experiment the only difference between the "units," the simulation runs, is the stream of random numbers used. In some circumstances it will make sense to block runs by keeping some or all of the random numbers constant. For example, in a queueing problem we have two sets of random variables, the arrival times and the service times. If these are generated by separate generators we can form blocks by holding either or both constant. In simulation parlance this is known as the method of *common random numbers*.

The classic problem with blocks is that they are perforce small. This does not apply in simulation, for we can make the blocks as large as we wish and still maintain complete homogeneity. There is thus no need to use incomplete block designs.

The literature is confused on the analysis of experiments involving common random numbers. If there are just two treatments we can take the difference in the responses and analyze these as independent samples. Let the responses be Y_1 and Y_2. Then

$$\text{var}(Y_1 - Y_2) = \text{var}(Y_1) + \text{var}(Y_2) - 2 \, \text{cov}(Y_1, Y_2)$$

Thus we will obtain a more precise estimate of the mean difference if $\text{corr}(Y_1, Y_2) > 0$. This is likely but need not always follow, as Wright and Ramsay (1979) demonstrate. This formula also demonstrates that common random numbers will be most effective when the treatments effect the mean response but not its variance.

The classic assumption for a randomized block design is that

$$\text{response} = \text{mean} + \text{treatment effect} + \text{block effect} + \text{error}$$

or

$$y_{ij} = \text{response to treatment } i \text{ on block } j$$
$$= \mu + \tau_i + b_j + \varepsilon_{ij}, \qquad \varepsilon_{ij} \sim N(0, \sigma^2)$$

If this holds, there is no difficulty in analyzing the experiment and producing estimates of treatment differences (and, more generally, treatment contrasts). Where blocks are created by common random numbers, this assumption amounts to assuming that the effect of changing the set of common random numbers is to change the response equally for all treatments. This appears to the author to usually be a tenable assumption. [Heikes et al. (1976) thought otherwise, and point out that a generalized least squares analysis is possible assuming correlation between the $\varepsilon_{.j}$ within each block.]

Two other worthwhile ideas from the design of experiments are *factorial experiments* and *response-surface* methodology. Where two or more factors may be varied in the design of a facility being studied by simulation, factorial experiments and particularly fractional factorials can help in producing an economical design. Where the aim is to minimize some cost, response surface modeling fits a "convenient" model to the estimated cost as a function of the variables and uses this to explore around the minimum cost conditions.

Schruben and Margolin (1978) report on a simulation study of a hospital specialized-care facility using both common random numbers and antithetic variates. Care is always needed in combining two variance reduction techniques, since all too often each defeats the other and no overall gain results. Their experimental design used six random number streams, and was unusual in that antithetic pairs were used at *different* points in a $2 \times 2 \times 2$ factorial experiment. The ideas incorporated in that study deserve wider consideration. Schruben and Margolin give proofs of variance reduction, but they do depend on inducing correlations of the correct signs, which we are not usually able to show theoretically.

The considerations of this chapter show that the design of simulation experiments is no easy matter. There always has to be a balance between the effort put in and the computer time saved. The author's experiments have almost all been sufficiently small but analytically intractable that the greatest benefits came from the judicious use of common random numbers and response surface modeling. With large but simple systems much more may be possible. It does seem that the topics of this chapter are widely ignored in published simulation studies, and when they are used the variance reductions gained have been modest, say 2–10. Reducing the cost from $3000 to $300 is clearly worthwhile; reducing $30 to $3 might not be with any realistic accounting of the simulator's time.

EXERCISES

5.1. Experiment with as many variance reduction techniques as you can think of to apply to the problem of evaluating $P(N > 2.5)$ for $N \sim N(0, 1)$.

5.2. Hammersley and Handscomb (1964) use the integration of $\phi(x) = (e^x - 1)/(e - 1)$ on $(0, 1)$ as a test problem of variance reduction techniques. Achieve as large a variance reduction as you can. (Hammersley and Handscomb achieved 4 million.)

5.3. Show that $\text{var}(\hat{\theta}_g)$ is minimized by $g \propto |\phi f|$ in importance sampling.

5.4. Show that for Bernoulli trials with probability $p \leqslant \frac{1}{2}$ of success the minimum achievable correlation is $-p/(1-p)$ and that this is achieved by counting events of probability $2p$.

5.5. Apply Cheng's method of antithetic variates to finding EW_{100} in a M/M/1 queue, and compare with Table 5.3.

5.6. An alternative way to write a control variates regression is

$$Z_i = \alpha + (\mathbf{w}_i - \bar{\mathbf{w}})\beta$$

Show that this gives an orthogonal design matrix, so under joint normality $\hat{\alpha}$ and $\hat{\beta}$ are independent with

$$\mathrm{var}(\hat{\alpha}) = \sigma_U^2/n$$

$$\mathrm{var}(\hat{\beta}) = \sigma_U^2(\mathbf{W}^T\mathbf{W})^{-1}$$

conditional on $(\mathbf{w}_i)$.
   Note that $\hat{\theta} = \hat{\alpha} + \bar{\mathbf{w}}\beta$, so

$$\mathrm{var}(\hat{\theta}) = \sigma_U^2\left[\frac{1}{n} + \bar{\mathbf{w}}(\mathbf{W}^T\mathbf{W})^{-1}\bar{\mathbf{w}}^T\right]$$

as before.

5.7. One way to perform antithetic runs is known as *seed switching*. Suppose the random number stream is produced by a maximal period multiplicative generator. Show that replacing the seed X_0 by $M - X_0$ produces the antithetic stream. (Thus no extra programming is required.) Is this possible with a mixed generator?

5.8. A stronger version of Theorem 5.3 is given by Wilson (1979). Suppose $U_1, \ldots, U_n$ have $U(0, 1)$ marginal distributions, and $g_j(u)$ are functions of finite variance. Then if $t = \Sigma_1^n g_j(u_j)$, the minimum variance of t is attained by cdfs in $\mathscr{H}$. Here a distribution $H \in \mathscr{H}$ is defined by functions $z_j : [0, 1] \to [0, 1]$ that are one-one, onto, and have derivative 1 except at a finite number of points. Then H is the cdf of the uniform distribution on the image of $[0, 1]$ under (z_j). Deduce this result for $n = 2$ from Theorem 5.3, and prove the general case.

5.9. Consider a renewal process $N(t)$. Show that applying antithetic variates to the lifetimes reduces $\mathrm{var}(\overline{N(t)})$ averaged over n runs (George, 1977).

5.10. Study Schruben and Margolin (1978) and consider how to apply their ideas to the queueing discipline example of Chapter 1.

5.11. An extension of antithetic variates termed *rotation sampling* for the simulation of Markov chains was introduced by Fishman (1983a, 1983b). He produces k parallel correlated runs by introducing correlation in the jumps made when the runs are in the same state. Suppose K runs are in state j. Then the next state is chosen by inversion with uniforms $U_r = (U + (r - 1)/K) \bmod 1, r = 1, \ldots, K$. This is applied at each step to each state in turn. Apply this to estimating the mean time to extinction of a branching process with $X_0 = 1$ and each individual divides into two with probability $p < \frac{1}{2}$ or dies at each generation.

CHAPTER 6 ·

Output Analysis

The analysis of simulation experiments that give one observation per run is relatively straightforward unless, as in Chapter 5, deliberate dependence was introduced in the experimental design. However, for simulations of systems evolving through time we may take many correlated observations per run. Suppose we are interested in the distribution of customer waiting time in a queueing system. Then the waits of each customer will be relevant data. Furthermore, we will expect the waiting times of successive arrivals to be correlated.

These problems arise also with the observation of actual queueing systems. However, because of the expense of observation, real systems are usually observed by sparse sampling (if at all), so the development of appropriate statistical analyses has been triggered by simulation experiments relatively recently.

The problems only arise for quantities defined on a system "in equilibrium" or in "steady state." That is, we have a strictly stationary process X_t on $(-\infty, \infty)$ or all the integers, and we are interested in aspects of the distribution of X_t for any fixed t. Alternatively, we can consider X_t starting at $t = 0$ and converging to an equilibrium process as $t \to \infty$. (The two ideas are equivalent in most examples, since the equilibrium process must be strictly stationary, and if a stationary process exists, there is usually a convergence theorem.) If we are interested in the *transient* behavior of the process, as in estimating $\omega = EW_{100}$ in Section 5.3, we will only have one observation per run. We now confine our attention to steady-state problems.

There are two approaches to simulating a steady-state problem. With n total observations we can take k runs of length m, where $mk \approx n$. The *terminating* simulation approach takes k large and hence m small. The *steady-state* approach takes a few long runs, perhaps only one. The terminating approach has a large number of independent replications and so is easy to analyze. Its results may however be biased by the problem of the initial transient. Only exceptionally are we able to sample a process in equilibrium. Normally we have to take a starting state at $t = 0$ and hope that the

142

distribution of X_t nears equilibrium well within the length of each run. We may well allow the process to "warm up" before observation by discarding data before $t = t_0 > 0$. The problem of detecting when a process has reached equilibrium is considered in Section 6.1.

The steady-state approach looks more promising in that only a small part of the data will need to be discarded before equilibrium is neared. However, it will be much more difficult to analyze, and in particular to produce reliable estimates of the variability of estimators and hence confidence intervals for parameters. Several different approaches are discussed in Sections 6.2 and 6.3.

Regenerative simulation provides an alternative analysis for either long or short runs of processes with regeneration points, although it is most appropriate to long runs. For this restricted class of processes it solves the problem of the initial transient and points a way to analyze correlated data. These apparently magical gains are somewhat illusory, for regenerative simulation will work well only when a large number of *tours* are observed, in which case the initial transient will be short and the correlation between observations will have a short range. Nevertheless it provides an interesting alternative, discussed in Section 6.4.

There are several instances in output analysis in which theoretical knowledge of the process can ease the statistical problems. The existence of regeneration points is one example. Another is the knowledge that the process is ergodic, without which basing results on one or a small number of runs is very dangerous. Unless the process is very well understood a minimum of $k = 3$ runs is advised to provide some check on estimates of variability produced internally to each run.

Suppose we are interested in $\theta = E[\phi(X_0)]$ under the steady-state distribution of X_0. Let $Y_1, Y_2, Y_3, \ldots$ be a series of observations on (X_t), and let $Z_i = \psi(Y_i)$, so $\theta = EZ_i$. Then most of our methods estimate $\hat{\theta} = \bar{Z}$, but differ in the way they estimate $\text{var}(\hat{\theta})$. The series (Z_i) is a stationary time series with autocorrelation sequence

$$\rho_s = \text{corr}(Z_t, Z_{t+s})$$

and variance $\sigma^2 = \text{var}(Z_i)$. Then

$$\text{var}(\bar{Z}) = n^{-2} \sum_{i,j} \text{cov}(Z_i, Z_j)$$

$$= \sigma^2 n^{-2} \sum_{i,j=1}^{n} \rho_{|i-j|}$$

$$= \sigma^2 n^{-1} \left[1 + 2 \sum_{1}^{n-1} \left(1 - \frac{s}{n} \right) \rho_s \right] \tag{1}$$

which will exceed σ^2 if all $\rho_s > 0$ (as is usual). Conversely, let $s^2 = \sum(Z_i - \bar{Z})^2/(n-1)$. Then

$$
\begin{aligned}
Es^2 &= (n-1)^{-1}E\sum(Z_i - \bar{Z})^2 \\
&= (n-1)^{-1}E\sum(Z_i^2 - 2Z_i\bar{Z} + \bar{Z})^2 \\
&= (n-1)^{-1}E[\sum(Z_i - \mu)^2 - n(\bar{Z} - \mu)^2] \\
&= (n-1)^{-1}[n\sigma^2 - n \, \text{var}\,\bar{Z}] \\
&= \sigma^2\left[1 - \frac{2}{n-1}\sum_{s=1}^{n-1}\left(1 - \frac{s}{n}\right)\rho_s\right]
\end{aligned}
$$

so s^2/n underestimates σ^2/n which underestimates var($\bar{Z}$), on average. A quantity often considered is

$$
\tau^2 = \lim_{n\to\infty} n\,\text{var}(\bar{Z}) = \sigma^2\left[1 + 2\sum_1^\infty \rho_s\right]
$$

Then var($\bar{Z}$) $\leqslant \tau^2/n$ if all correlations $\rho_s > 0$, and the relative difference is likely to be small for large n. We can also express τ^2 in terms of the spectral density f of (Z_i). This is a positive function f on $[0, \pi)$, defined so that

$$
\sigma^2\rho_s = 2\int_0^\pi \cos s\omega \, f(\omega)d\omega
$$

Then

$$
f(\omega) = \frac{\sigma^2}{2\pi}\sum_{-\infty}^\infty \rho_s \cos s\omega
$$

so

$$
2\pi f(0) = \sigma^2\sum_{-\infty}^\infty \rho_s = \sigma^2\left[1 + 2\sum_1^\infty \rho_s\right] = \tau^2
$$

This formula will be taken up in Section 6.3.

Moran (1975) considers the traditional methods to estimate var($\bar{Z}$). One is to substitute estimators of ρ_s and σ^2 in (1). Hannan (1957) proposed

$$
\hat{V}_1 = \frac{n}{(n-l)(n-l+1)}\sum_{|s|<l}\left(1 - \frac{|s|}{n}\right)(c_s - \bar{Z}^2)
$$

where

$$c_s = \sum \frac{Z_i Z_{i+s}}{n-s}$$

and l is chosen so that $\rho_s \approx 0$ for $s \geqslant l$. Then Moran shows .

$$E\hat{V}_1 = \text{var}(\bar{X}) - \frac{2\sigma^2 n}{(n-l)(n-l+1)} \sum_{s=l}^{n-1} \left(1 - \frac{s}{n}\right) \rho_s$$

so $\hat{V}_1$ is approximately unbiased if its assumption holds. However, if l is appreciable compared to n then $\hat{V}_1$ is quite variable. An alternative approach is to estimate ρ_s parametrically, as discussed in Section 6.3.

The other traditional approach is "batching" or the method of batch means. The data $(Z_1, \ldots, Z_n)$ are divided into k batches of length m with means (B_i), so

$$B_i = m^{-1}[Z_{(i-1)m+1} + \cdots + Z_{im}]$$

Then we hope $(B_1, \ldots, B_k)$ is relatively uncorrelated and that

$$\hat{V}_2 = \frac{1}{k(k-1)} \sum (B_i - \bar{B})^2$$

is a better estimator of $\text{var}(\bar{Z}) = \text{var}(\bar{B})$. This is discussed in more detail in Section 6.2. An important side effect is that we may expect the B_i to be nearly normally distributed. The final advantage of batching is that it reduces the volume of data to be stored and manipulated. Some simulators and simulation systems routinely batch all data.

Much of the work on output analysis gives only asymptotic results, as n or m and/or k tend to infinity. This is no great disadvantage as exact results would depend heavily on the system under study. The asymptotic results provide a fairly general approximation. Since (Z_i) are dependent we need limit results for dependent sequences. Many different results are available, of which some of the most useful appear to be those depending on ϕ-mixing (Billingsley, 1968). Let $\phi(n)$ be a positive sequence decreasing to zero. We will require that $\sum \phi(n)^{1/2} < \infty$. Then (Z_i) is ϕ-mixing if

$$|P(B|A) - P(B)| \leqslant \phi(r)$$

whenever A depends on $(Z_1, \ldots, Z_i)$ and B on $(Z_{i+r+1}, \ldots)$. Most queueing systems do satisfy this condition, which essentially rules out long-range dependence. Let $S_n = Z_1 + \cdots + Z_n$. Then, if $[nt]$ denotes the integer part of nt,

$$W_n(t) = (S_{[nt]} - [nt]\theta)/\tau\sqrt{n}, \qquad 0 \leqslant t \leqslant 1$$

converges weakly to Brownian motion, $W_n \Rightarrow W$. [This implies that $W_n(t) \Rightarrow W(t)$ for each t, as well as stronger results such as max $W_n \Rightarrow$ max W.] The definition of W_n includes the unknown θ, so for some purposes we replace θ by $Z_n = S_n/n$, to get

$$B_n(t) = (S_{[nt]} - [nt]Z_n)/\tau\sqrt{n}$$

Note that $B_n(0) = B_n(1) = 0$. We can then show that $B_n \Rightarrow B_0$, the Brownian bridge process. (See Exercise 4.7.) The process $-B_n$ is termed a *standardized time series* by Schruben (1982, 1983). These results are used to give approximate distributions for various test statistics below.

6.1. THE INITIAL TRANSIENT

Throughout this section we assume that X_0 is fixed, and $Z_1, Z_2, \ldots$ are observed, with (Z_i) converging in distribution and L_1 to Z_∞, so $\theta = EZ_\infty$.

Having observed $Z_1, \ldots, Z_n$ we delete $Z_1, \ldots, Z_d$ and estimate θ by

$$\theta_d = (n - d)^{-1} \sum_{i=d+1}^{n} Z_i$$

We hope that $|E\theta_d - \theta|$ decreases as d increases. However var(θ_d) may increase with d, so it is by no means clear that if we are only interested in θ we should delete any of the series. It has been suggested in the literature that d be chosen to minimize the mean square error of θ_d. The choice of d will depend on n and Z_0 and may be $d = 0$, as shown by Blomquist (1970) for the average waiting time in a M/M/1 queue for large n. However, we will be more interested in a confidence interval for θ, and experience has shown that deletion leads to possibly wider but much more reliable confidence intervals.

A wide variety of ad hoc tests for "steady state" and algorithms for deletion have been proposed. Wilson and Pritsker (1978) and Gafarian et al. (1978) survey some of these proposals. None have been found reliable in subsequent simulation experiments, and attention has turned to the more formal tests described later. These are all tests of equality of $\mu_i = EZ_i$. We know $\mu_i \to \theta$, and we may know that μ_i increases or decreases, for example, in certain queueing systems starting up from their empty state. The tests implicitly assume $\text{var}(Z_i) = \sigma^2$, although this may not be justified. They would therefore *not* detect the initial transient in the $AR(1)$ process

$$Z_i = \alpha Z_{i-1} + \varepsilon_i, \qquad i > 0, \qquad Z_0 = 0$$

since $EZ_i \equiv 0$ but

$$\text{var}(Z_i) = \alpha^2 \text{var}(Z_{i-1}) + \sigma_\varepsilon^2$$

so

$$\text{var}(Z_i) = \sigma_\varepsilon^2(1 + \alpha^2 + \cdots + \alpha^{2i-2}) = \frac{\sigma_\varepsilon^2(1 - \alpha^{2i})}{1 - \alpha^2}$$

increases to $\text{var}(Z_\infty)$.

The simplest procedures are graphical. We can estimate μ_i by taking r replications of $(Z_1, \ldots, Z_n)$ and letting $\hat{\mu}_i$ be the average of the observations of Z_i. Unless r is very large these estimates will be too variable, but they can be smoothed before plotting. A simple way to smooth is to use a moving average of the form

$$\tilde{\mu}_i = (2b + 1)^{-1} \sum_{t=-b}^{b} \hat{\mu}_{i+t}$$

with suitable adjustments near 1 and n. Then b is chosen to obtain a smooth picture. (More sophisticated smoothers can be used, including monotone regression if μ_i is known to be monotone.) The value of d is then chosen as the point at which $\tilde{\mu}_i$ appears to have converged. Figure 6.1 illustrates the procedure. Welch (1983) gives other examples. In general, the choice of d is not easy.

Automatic procedures for detecting and deleting an initial transient are based on significance tests of $\mu_1 = \mu_2 = \cdots = \mu_n = \theta$. The experience of cusum techniques in quality control suggests that a gradually drifting mean (such as seen in Fig. 6.1a viewed from right to left) is best detected from cumulative sums of the form

$$S_k' = \sum_{i=k+1}^{n} (Z_i - \theta), \qquad k = n - 1, \ldots, 0$$

Unfortunately, θ is unknown, so we substitute $\hat{\theta} = \bar{Z}$. Then

$$S_k' = -(S_k - k\bar{Z})$$

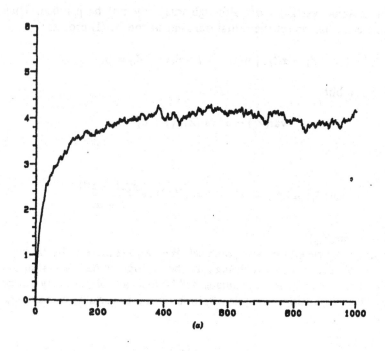

(a)

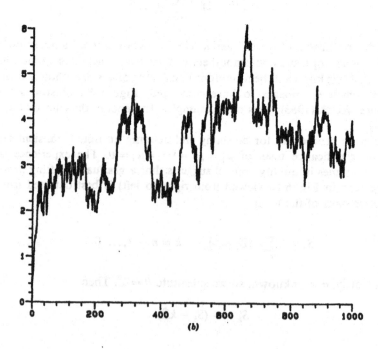

(b)

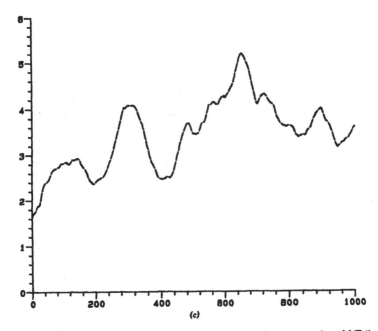

Figure 6.1. Plots of EW_i vs i for the waiting time W_i of the ith customer in a M/D/1 queue starting from empty. The queue had unit arrival rate and $\rho = 0.9$. Parts (a) and (b) show unsmoothed averages of 1000 and 25 runs, respectively. Part (c) is (b) smoothed by a moving average with $b = 25$. It is clear that the choice of how much to delete is more difficult from (b) or (c) than from (a).

so we may as well work with the sums $\sum (Z_i - \bar{Z})$, for which we have a limit theorem via the B_n process. If there *is* a drift then B_n will have large fluctuations, so we can base a significance test on any statistic indicating large fluctuations in B_n or B_0. Analogous problems in other branches of statistics with the same limit process suggest

$$CM = \int_0^1 B_n(t)^2 dt$$

the Cramér–von Mises statistic favored by Heidelberger and Welch (1983). Schruben (1982) gives a one-sided test for μ_i increasing

$$S = B_n(t^*)^2 / t^*(1 - t^*)$$

where min $B_n(t)$ is attained at t^*. The weak convergence theory gives asymptotic distributions for *CM* and *S*. For *CM* we have 95% and 99% points of 0.46 and 0.74, whereas for *S*, Schruben shows $S \sim \chi_3^2$. To apply either test we need a consistent estimator of τ (for which the asymptotic distribution remains valid); such estimators are discussed in the next two sections. See also Schruben et al. (1983).

An automatic procedure will apply such a test to a series of deletion points *d*. Heidelberger and Welch (1981a, 1983) recommend deleting 0%, 10%, 20%, 30%, 40%, and 50% in turn, stopping when the remaining series passes the *CM* test at the 5% level. If all six tests are failed a longer series is needed, so *n* is increased and the process is repeated. When a satisfactorily stationary series is found the width of a confidence interval for θ is computed, and *n* is increased further if necessary to achieve a width less than a prescribed limit. (The whole procedure is in fact applied to batches rather than the original data.)

The multiple use of significance tests in these automatic procedures is problematical. In general some supervision will be advisable, and knowledge of the process being studied can be very helpful. It would seem advisable to delete too much, and produce a safe confidence interval for θ, than to delete too little and experience bias.

6.2. BATCHING

The use of batches is a time-honored way to cope with correlation within a series $(Z_1, \ldots, Z_n)$. Suppose this is divided into *k* successive batches of length *m* with batch means $B_1, \ldots, B_k$. The correlations between (B_i) should be less than those between (Z_i). Now

$$\text{var}(B_i) = m^{-2} \sum_{s,t=1}^{m} \rho_{|s-t|}\sigma^2$$

$$= m^{-1}\sigma^2 \left[1 + 2\sum_{1}^{m-1}\left(1 - \frac{s}{m}\right)\rho_s\right]$$

$$\text{cov}(B_i, B_j) = m^{-2}\sigma^2 \sum_{s,t=1}^{m} \rho_{|(j-i)m+t-s|}$$

For example, suppose (Z_i) is a $AR(1)$ process, so $\rho_s = \alpha^{|s|}$ for $|\alpha| < 1$. Then for $i < j$

$$\text{cov}(B_i, B_j) = m^{-2}\sigma^2 \frac{\alpha^{1-m} - \alpha^m}{(1-\alpha)^2} \alpha^{m(j-i)}$$

so (B_i) has a geometrically decaying autocorrelation at rate α^m. Let

$$V_2 = \frac{1}{k(k-1)}(B_i - \bar{B})^2$$

be the batching estimator of $\text{var}(\bar{Z})$. Then Moran (1975) shows that

$$E\hat{V}_2 = \text{var}(\bar{Z}) - \frac{2\sigma^2}{m(k-1)}\left[\sum_1^{m-1}\frac{n-m}{nm}s\rho_s + \sum_m^{n-1}\left(1 - \frac{s}{n}\right)\rho_s\right]$$

Thus $\hat{V}_2$ will be effectively unbiased provided n is large and $\rho_s \approx 0$ for $s \geq m$. For the $AR(1)$ process we obtain

$$E\hat{V}_2 = \text{var}(\bar{Z}) - \frac{2\sigma^2\alpha}{m(k-1)(1-\alpha)^2}\left[\frac{1-\alpha^m}{m} - \frac{1-\alpha^n}{n}\right]$$

with

$$\text{var}(\bar{Z}) = \frac{\sigma^2}{n}\left[\frac{1+\alpha}{1-\alpha} - \frac{2\alpha}{n(1-\alpha)^2}(1-\alpha^n)\right]$$

and $\tau^2 = \sigma^2(1+\alpha)/(1-\alpha)$. Thus the relative bias is small provided $m^{-1}(1-\alpha^m)$ is small. Tables 6.1 and 6.2 compare $\hat{V}_1$ and $\hat{V}_2$ for the $AR(1)$ process. Note that

$$\text{var}(\hat{V}_2) \approx \frac{2\,\text{var}(\bar{Z})^2}{k-1}$$

Table 6.1. The Performance of $\hat{V}_1$ for an $AR(1)$ Process with $\sigma^2 = 1$

	$\text{var}(\bar{Z})$	$\text{bias}(\hat{V}_1)$	$\text{var}(\hat{V}_1)$	$\text{cv}(\hat{V}_1)$
$n = 100, \alpha = 0.5$				
$l = 10$	2.98×10^{-2}	-0.14%	3.8×10^{-4}	0.64
$n = 1000, \alpha = 0.5$				
$l = 10$	3.00×10^{-3}	-0.13%	3.8×10^{-7}	0.20
$n = 1000, \alpha = 0.9$				
$l = 10$	1.90×10^{-2}	-37%	6.4×10^{-6}	0.21
$l = 20$	1.90×10^{-2}	-13%	1.9×10^{-5}	0.26
$l = 50$	1.90×10^{-2}	-0.56%	5.9×10^{-5}	0.41

Table 6.2. Performance of Batching, with k Batches of Length m Each, for an $AR(1)$ Process with $\sigma^2 = 1$

	corr(B_i, B_{i+1})	bias$(\hat{V}_2)$	var$(\hat{V}_2)$	cv$(\hat{V}_2)$
$n = 100, \alpha = 0.5$				
$k = 2$	1.37%	-2.7%	2.1×10^{-3}	1.5
$k = 5$	3.57%	-6.7%	3.9×10^{-4}	0.71
$k = 10$	7.67%	-13.4%	1.6×10^{-4}	0.48
$n = 1000, \alpha = 0.5$				
$k = 10$	0.68%	-1.3%	1.9×10^{-6}	0.48
$k = 50$	3.57%	-6.7%	3.8×10^{-7}	0.21
$n = 1000, \alpha = 0.9$				
$k = 5$	2.49%	-4.7%	1.9×10^{-4}	0.76
$k = 10$	5.2%	-9.5%	6.5×10^{-5}	0.48

A side effect of batching will be that each B_i is approximately normally distributed. [Indeed, the central limit theorem for the W_n process will give us more, asymptotic joint normality of $(B_1, \ldots, B_k)$ as $m \to \infty$. Brillinger (1973) gives a more direct proof of asymptotic normality.] This and independence gives a χ^2 distribution of $\hat{V}_2$ and the stated variance.

There is necessarily a compromise between choosing m large to achieve negligible correlation between batches and having k large enough to obtain a precise estimate of $v^2 = \text{var}(B_i)$. We will use a $(1 - \alpha)$ confidence interval for θ of the form

$$(\bar{B} - t_{k-1,\alpha}\sqrt{\hat{V}_2}, \bar{B} + t_{k-1,\alpha}\sqrt{\hat{V}_2})$$

This is valid provided m is large enough to make (B_i) approximately independent and normally distributed. The added variability due to $\hat{V}_2$ will be appreciable for $k \leqslant 30$. [Schmeiser (1982) calculates the effect of choosing k too small.] Since $\bar{B} = \bar{Z}$ independently of k, the choice of k too small gives a confidence interval that is unnecessarily wide, whereas choosing k too large causes $\hat{V}_2$ to be biased downward and so gives a confidence interval that is optimistically short.

The recommendations for chosing k in the literature boil down to testing for serial dependence in (B_i). For example, it has been suggested to test that the lag-one correlation is less than 0.05. A related test (Fishman, 1978b; Kleijnen et al., 1982) is von Neumann's test, which is a test of $\rho_1 = 0$, applied to (B_i). The problem is that large values of k will be necessary to establish $\rho_1 \neq 0$! This is another problem in which *a priori* theoretical knowledge can be very helpful in deciding on a minimum for m.

The method of "standardized time series" can be used to produce estimators of τ^2 (Schruben, 1983). Whereas $\hat{V}_2$ is based on the between-batch variation, this method uses the within-batch variability. Consider first $X_1, \ldots, X_m$, the contents of a single batch. This gives rise to a standardized time series

$$B_m(t) = [(X_1 - \bar{X}) + \cdots + (X_{[mt]} - \bar{X})]/\tau\sqrt{m}$$

and $B_m \Rightarrow B_0$ as $m \to \infty$. We can use the known variability of B_0 to estimate τ^2. For example, let

$$A = \tau\sqrt{m} \sum_{j=1}^{m} B_m(j/m) = \sum_{j=1}^{m} \left[\sum_{i=1}^{j} (X_i - \bar{X}) \right]$$

Then $A \sim N(0, \tau^2 m(m^2 - 1)/12)$ asymptotically, so

$$12A^2/m(m^2 - 1)$$

estimates τ^2. If this is applied to block i to obtain A_i, then

$$\hat{V}_3 = (nk)^{-1} \sum_{1}^{k} A_i^2 \times [12/m(m^2 - 1)]$$

is asymptotically as $m \to \infty$, k fixed, an estimator of τ^2/n, with $\hat{V}_3 \times kn\tau^{-2} \sim \chi_k^2$. A second measure of the variability of B_0 is

$$B_0(t^*)^2/t^*(1 - t^*), \qquad B_0(t^*) = \max B_0(t)$$

from which we get

$$B = m \left[\sum_{j=1}^{I} (X_i - \bar{X}) \right]^2 \bigg/ I(m - I)$$

with I attaining $\max_j \sum_{1}^{j} (X_i - \bar{X})$ and

$$\hat{V}_4 = \sum_{1}^{k} \frac{B_i}{3kn}$$

estimates τ^2/n, with $\hat{V}_4 \times 3nk\tau^{-2} \sim \chi_{3k}^2$.

As $m \to \infty$ both $\hat{V}_3$ and $\hat{V}_4$ are asymptotically independent of $\hat{V}_2$, so we can combine both within- and between-batch information. Schruben (1983) reports on some experiments with $\hat{V}_2, \hat{V}_3$, and $\hat{V}_4$ and combinations, which generally favor $\hat{V}_4$. However, he takes m very large and k small and so biases the comparison against $\hat{V}_2$.

Tables 6.1–6.3 report a simulation experiment with $\hat{V}_1$–$\hat{V}_4$ for an $AR(1)$ process of unit variance. Points to note are the very small bias of $\hat{V}_1$, the need for m to be quite large for a reasonable bias for $\hat{V}_2$, and the consistent underestimation by $\hat{V}_3$ and particularly $\hat{V}_4$ unless m is very large. Although $\hat{V}_1$ has been neglected in the simulation literature, it turned out to be best in this experiment. This example shows that the compromise between bias and variance is serious for all estimators. In some case the minimum mean square error is achieved with 20% or more bias, but this will give severely optimistic confidence intervals.

For $\hat{V}_3$ we can compute the bias explicitly. We know

$$E\hat{V}_3 = (nk)^{-1}\sum EA_i^2 \times 12/m(m^2 - 1)$$
$$= n^{-1} EA^2 \times 12/m(m^2 - 1)$$

Now

$$A = \sum_{j=1}^{m}\sum_{i=1}^{j}(X_i - X) = \sum X_i\left(\frac{m+1}{2} - i\right)$$

Table 6.3. Performance of the Within-Batch Estimators $\hat{V}_3$ and $\hat{V}_4$ with k Batches for an $AR(1)$ Process with $\sigma^2 = 1$

	$\hat{V}_3$		$\hat{V}_4$	
	Mean	Variance	Mean	Variance
$n = 100, \alpha = 0.5$				
$k = 2$	2.76×10^{-2}	8.5×10^{-4}	1.6×10^{-2}	1.3×10^{-4}
$k = 5$	2.41×10^{-2}	2.6×10^{-4}	0.95×10^{-2}	3.3×10^{-5}
$k = 10$	1.89×10^{-2}	7.1×10^{-5}	0.60×10^{-2}	7.1×10^{-6}
$n = 1000, \alpha = 0.5$				
$k = 10$	2.88×10^{-3}	1.7×10^{-6}	1.9×10^{-3}	4.5×10^{-7}
$k = 50$	2.41×10^{-3}	2.9×10^{-7}	1.1×10^{-3}	3.5×10^{-8}
$n = 1000, \alpha = 0.9$				
$k = 5$	1.63×10^{-2}	9.5×10^{-5}	0.86×10^{-3}	2.5×10^{-5}
$k = 10$	1.39×10^{-2}	3.7×10^{-5}	0.54×10^{-3}	5.5×10^{-6}

so

$$EA^2 = \sum_{i,j} \text{cov}(X_i, X_j)\left(\frac{m+1}{2} - i\right)\left(\frac{m+1}{2} - j\right)$$

For an *independent* sequence

$$EA^2 = \sum_i \sigma^2 \left(\frac{m+1}{2} - i\right)^2 = \frac{\sigma^2}{12} m(m^2 - 1)$$

showing that $\hat{V}_3$ is unbiased. In general

$$E\hat{V}_3 = \frac{\sigma^2}{n} \sum_{i,j} \rho_{|i-j|}\left(\frac{m+1}{2} - i\right)\left(\frac{m+1}{2} - j\right) \times \frac{12}{m(m^2 - 1)}$$

$$= \frac{\sigma^2}{n} \sum_{s=0}^{m-1} c_s \rho_s$$

for constants $c_s(m)$, that can be computed. The mean values of Table 6.3 were checked against this formula and confirm the serious bias of $\hat{V}_3$ for small m. For both $\hat{V}_3$ and $\hat{V}_4$ we appear to need very large batches for the asymptotic results to be a reasonable approximation, much larger batches than are necessary for $\hat{V}_2$, the between-batch estimator.

Our recommendation is to use $\hat{V}_1$, or $\hat{V}_2$ with m large enough to achieve correlation between adjacent batches of less than 5%.

6.3. TIME-SERIES METHODS

We have already seen that under mild conditions we can suppose that $\bar{Z} \sim N(\theta, \tau^2/n)$. Time-series methods provide yet another way to estimate τ^2. Rather than estimate (ρ_s) nonparametrically as in $\hat{V}_1$, we can fit a model to the data $(Z_1, \ldots, Z_n)$ and use its value of $\tau^2 = 2\pi f(0)$. A common example is the $AR(p)$ model [e.g., Fishman (1971)]. Then

$$(Z_i - \theta) = \sum_1^p \alpha_j(Z_{i-j} - \theta) + \varepsilon_i, \qquad \varepsilon_i \sim N(0, \sigma_\varepsilon^2)$$

for which

$$f(\omega) = \sigma_\varepsilon^2/2\pi|1 - \sum \alpha_j e^{-ij\omega}|^2$$

so

$$\tau^2 = 2\pi f(0) = \sigma_\varepsilon^2 \Big/ \left(1 - \sum_1^p \alpha_j\right)^2$$

Thus an estimate of var($\bar{Z}$) is

$$\hat{V}_s = \hat{\sigma}_\varepsilon^2/n \left(1 - \sum_1^p \hat{\alpha}_j\right)^2$$

Standard time-series methods can be used to select p and estimate the parameters $\alpha_1, ..., \alpha_p, \sigma_\varepsilon^2$ (Priestley, 1981, Section 5.4). If no time-series package is available, the simplest method is to regress $(Z_{p+1}, ..., Z_n)$ on $(Z_{p+1-i}, ..., Z_{n-i})$, $i = 1, ..., p$. Then the regression coefficients estimate $\alpha_1, ..., \alpha_p$ and the residual mean square estimates σ_ε^2. This can be tried for various values of p and Akaike's *AIC* criterion

$$\overline{AIC}(p) = n \ln \hat{\sigma}_\varepsilon^2(p) + 2p$$

minimized by the choice of p.

We can also fit $ARMA(p, q)$ models of the form

$$(Z_i - \theta) = \sum_1^p \alpha_j(Z_{i-j}^i - \theta) + \sum_0^q \beta_j \varepsilon_{i-j}$$

with

$$\tau^2 = \sigma_\varepsilon^2 \left(1 + \sum_1^q \beta_j\right)^2 \Big/ \left(1 - \sum_1^p \alpha_j\right)^2$$

but these have been used infrequently, since although fewer parameters may be needed parameter estimation is more difficult computationally.

It is worth noting that since we are effectively estimating $f(0)$, the method previously described is one version of autoregressive spectral estimation. Further methods are described by Priestley (1981, Sections 7.8 and 7.9).

The traditional methods for estimating $f(\omega)$ are based on smoothing the periodogram and are described by Priestley (1981, Sections 6.2, 7.4–7.6). Duket and Pritsker (1978) and Heidelberger and Welch (1981a, 1981b) discuss the use of these nonparametric methods for the estimation of $f(0)$. The problem with the usual smoothing methods is that they give a good idea of *shape* of $f( )$ but underestimate peaks and overestimate troughs. This is a particular problem here since almost all the processes (Z_i) of interest will have $f(\omega)$ decreasing rapidly for small ω, so $\tau^2 = 2\pi f(0)$ will be underestimated. Heidelberger and Welch propose special smoothing methods and advocate fitting a local quadratic at $\omega = 0$ to $\ln f(\ )$. However, the most promising idea is to use *prewhitening* (Priestley, 1981, pp. 556–557). This fits an $AR(p)$ process for small p, converting $(Z_1, \ldots, Z_n)$ to $(\hat{\varepsilon}_1, \ldots, \hat{\varepsilon}_n)$. Then

$$f(\omega) = \frac{f_{\hat{\varepsilon}}(\omega)}{\left| 1 - \sum_1^p \hat{\alpha}_j e^{-ij\omega} \right|^2}$$

and $f_{\hat{\varepsilon}}(\omega)$ should be relatively flat so $f_{\hat{\varepsilon}}(0)$ can be estimated reliably from $(\hat{\varepsilon}_1, \ldots, \hat{\varepsilon}_n)$. Then

$$\hat{f}(0) = \hat{f}_{\hat{\varepsilon}}(0)/(1 - \sum \hat{\alpha}_j)^2$$

It should be clear that time-series methods need a good deal of expertise to be used. Except for the straight periodogram smoothing methods it is difficult to estimate $\mathrm{var}(\hat{\tau}^2)$. On the positive side the methods of this section do genuinely take correlation into account rather than rely on dubious assumptions that it has been circumvented. They are probably the best methods for estimating τ^2 for an expert user with access to a good time-series package.

6.4. REGENERATIVE SIMULATION

"Regenerative simulation" is a term coined by Iglehart and his co-workers for a method of output analysis for regenerative processes. It stems from a remark of Cox and Smith (1961, p. 136)

> In many systems ... the process falls naturally into sections of unequal length, behavior in different sections being independent.... It follows that the equilibrium properties of the system can be derived from those of tours, a tour starting with an arrival at an empty system and ending the next time the system is again empty.

Some early users of this idea were Kabak (1968) and Fishman (1973, 1974), before it was taken up in a series of papers by Crane and Iglehart (1974a, 1974b, 1975a, 1975b), and Iglehart (1975, 1976, 1977). Crane and Lemoine (1977) and Iglehart and Shedler(1980) both present the subsequent developments in published lecture notes.

It is easiest to explain the analysis by an example. We return to Table 1.3, a simulation of an M/D/3 queue. Asterisks mark customers who arrived at an empty queue, so there were seven completed tours and one incomplete one in the 200 observations. We discard the incomplete tour, giving

Length of Tour	Total Wait
1	0
28	4.692
12	1.663
8	0.387
26	5.894
41 ,	15.922
37	22.787

We are interested in the mean waiting time in equilibrium, θ. An obvious estimator of θ is the total waiting time divided by the total number of customers served,

$$\hat{\theta} = 51.35/153 = 0.3356$$

(Note this is the total over *complete* tours; discarding the incomplete tour is analogous to discarding an initial transient.) This is a ratio of random quantities, so it will not necessarily be unbiased. To estimate its variance, we compute

$$\hat{\sigma}^2 = \frac{1}{n_t - 1} \sum (y_i - \hat{\theta} x_i)^2 = 25.56$$

for the n_t tours with totalled observations (x_i, y_i). Then

$$\text{s.e.}(\hat{\theta}) \approx \hat{\sigma}/\bar{x}\sqrt{n_t} = 0.087$$

which will be subsequently justified.

Another way to estimate means and variances is to use the jacknife (Efron, 1982). Let $\bar{y}_{(i)}$ denote the mean of (y_i) omitting tour i, and similarly for $\bar{x}_{(i)}$. Let $\hat{\theta}_{(i)} = \bar{y}_{(i)}/\bar{x}_{(i)}$. Then

$$\tilde{\theta} = n_t \hat{\theta} - (n_t - 1)\bar{\theta} \quad \text{where} \quad \bar{\theta} = n_t^{-1} \sum \hat{\theta}_{(i)}$$

is the jacknife estimator of θ, and

$$\tilde{s}^2 = \frac{n_t - 1}{n_t} \sum (\tilde{\theta} - \theta_{(i)})^2$$

is the jacknife estimator of $\text{var}(\hat{\theta})$ or $\text{var}(\tilde{\theta})$. From Table 1.3 $\tilde{\theta} = 0.3476$ and $\tilde{s} = 0.096$. The jacknife estimates are generally preferred, for reasons discussed later.

To underpin this analysis some theory has been developed. A *regenerative process* has a sequence of regeneration points $T_1, T_2, \ldots$ that are random stopping times at which the future of the process is independent of its past. The *tours* between regeneration points are then independent, and the sequence of regeneration points is a renewal process. Let (X_t) be the sequence of tour lengths giving rise to the renewal process $N(t)$, and let Y_r be the total observation on the rth trip. Then (X_r, Y_r) are independent and identically distributed. Let

$$Y(t) = \sum_1^{N(t)} Y_r$$

which is the total observation on observed tours to date. We assume both $EX_1 < \infty$ and $E|Y_1| < \infty$. Then

$$P(Y(t)/t \rightarrow EY_1/EX_1) = 1$$

$$EY(t)/t \rightarrow EY_1/EX_1$$

(Ross, 1970, Theorem 3.16). Some further technicalities give us the same results for $Y'(t)$, the total observation by time t. Thus we identify $\theta = EY_1/EX_1$ as the mean observation in equilibrium. Note that this is the ratio of means, not the mean of the ratio as one might expect.

Now consider $\hat{\theta} = \bar{Y}/\bar{X}$ [averaged over $N(t)$ tours]. We know $P(N(t) \rightarrow \infty) = 1$ from renewal theory, so $P(\hat{\theta} \rightarrow \theta) = 1$ by the strong law of large numbers. Thus $\hat{\theta}$ is a strongly consistent estimator of θ, but it will in general be biased. Let $V_j = Y_j - \theta X_j$. Then (V_j) are independent with mean zero, and by the central limit theorem we may assume that $\bar{V} \sim N(0, \sigma^2/n_t)$ approximately, where $\sigma^2 = \text{var}(V_1)$. Thus

$$\sqrt{n_t}(\bar{Y} - \theta\bar{X})/\sigma \sim N(0, 1)$$

so

$$\sqrt{n_t}(\hat\theta - \theta)/(\sigma/\bar{X}) \sim N(0, 1)$$

or

$$\hat\theta \sim N(\theta, \sigma^2/n_t\bar{X}^2), \text{approximately.}$$

This gives the approximate $(1 - \alpha)$-confidence interval

$$(\hat\theta - k_\alpha\sigma/\bar{X}\sqrt{n_t}, \hat\theta + k_\alpha\sigma/\bar{X}\sqrt{n_t})$$

We estimate $\sigma^2 = \text{var}(Y_1 - \theta X_1)$ by the sample variance of $(Y_i - \hat\theta X_i)$ to get $\hat\sigma^2$.

This procedure is a standard way to assess the variability of ratio estimators in sampling theory. It does depend on $\bar{X}$ being essentially constant and so needs n_t large. An alternative justification would be an analysis conditional on (X_i). If $E(Y_i|X_i) \approx \theta X_i$ and $\text{var}(Y_i|X_i) \approx \sigma^2$ then we obtain the same approximate confidence interval conditionally and hence unconditionally. However, neither assumption is likely to hold, particularly that of the variance, which we might expect to increase with X_i. A more plausible assumption is $\text{var}(Y_i|X_i) = \gamma X_i$, which gives $\text{var}(\hat\theta) \approx \gamma/\bar{X}$. Applied to our example we get s.e.$(\hat\theta) \approx 0.09$, but it is clear that $E(Y_i|X_i) = \theta X_i$ is also violated.

The original rationale behind Quenouille's introduction of the jacknife was that if $E\hat\theta - \theta = O(n^{-1})$ then the bias of $\tilde\theta$ would be $O(n^{-2})$. Efron (1982) discusses the properties of $\tilde\theta$ and $\tilde{s}^2$, and shows that $\tilde{s}^2$ is usually conservative in the sense that $E\tilde{s}^2 \geqslant \text{var}(\tilde\theta)$. Simulation experiments have shown the jacknife estimator $\tilde{s}^2$ to generally be more reliable than $\hat\sigma^2/n_t\bar{X}^2$ as an estimator of $\text{var}(\hat\theta)$. Thus we would recommend the use of $\tilde{s}^2$. The difference between $\hat\theta$ and $\tilde\theta$ is almost always small, so the choice between them is not important. Yet another alternative, the bootstrap, is discussed in Section 7.1.

Each of these analyses is really a large-sample result in the sense of a large number of tours, for attempts to justify confidence intervals of the form

$$(\tilde\theta - t_\alpha\tilde{s}, \tilde\theta + t_\alpha\tilde{s})$$

for t_α the $(1 - \alpha/2)$ point of a t distribution have been largely unsuccessful (Efron, 1982, p. 14). Thus valid inferences depend on $\tilde{s}^2$ being an accurate estimator and need n_t large. A large number of tours implies that the correlation of the original process is of short range compared to the length of observation.

Regenerative simulation has been applied to a wide range of problems in the references already cited and in Heidelberger and Iglehart (1979), Iglehart and Lewis (1979), Iglehart and Shedler (1978, 1981, 1983a, 1983b), Iglehart and Stone (1983), Lavenberg and Sauer (1977), Meketon and Heidelberger (1982), Seila (1982) and Shedler and Southard (1982). Some of these refer to identifying the regenerative points or choosing between regenerative points (as in Markov chains). Others refer to the use of statistical procedures that are complicated by the random number of tours. One can also consider an approximate regenerative analysis as in Crane and Iglehart (1975b) and Gunter and Wolff (1980).

Meketon and Heidelberger (1982) consider the problem of the incomplete tour at time t. By the waiting-time paradox this is no ordinary tour but is likely to be longer than expectation. They show that

$$E\hat\theta = \theta + c/t + O(1/t^2)$$

for

$$c = EY_1\left[\frac{\operatorname{var} X_1}{E(X_1)^2} - \frac{\operatorname{cov}(Y_1, X_1)}{E(X_1)E(Y_1)}\right]$$

under technical conditions, whereas if observation is continued to the end of the incomplete tour,

$$E\hat\theta = \theta + O(1/t^2)$$

This has a similar empirical performance to jacknifing, with a simpler analysis but a more costly simulation.

6.5. A CASE STUDY

Queueing systems are normally studied with identical servers. Suppose we have one slow and one fast server. Are there any circumstances in which it is preferable to run a single-server queue with just the fast server? We will only consider first-in–first-out (FIFO) systems for which it is clear that adding an additional server (however slow) will reduce the waiting time for each customer. However, this could be balanced by the increased service time for those customers unfortunate enough to be allocated to the slow server.

Rubinovitch (1985) considered this problem for mean delay time in queues with Poisson arrivals of rate λ and exponential service rates $\mu_1 > \mu_2$. Then both one- and two-server queues are Markov processes which can be studied analytically. Figure 6.2 illustrates Rubinovitch's results. At high traffic intensities the slow server is worthwhile unless it is extremely slow. At low intensities most of the customers find both servers free. If they cannot distinguish between the servers they will choose one at random, in which case only slightly slower servers are acceptable. If the servers are labeled the fast server will be chosen if it is free. Then a server up to twice as slow is acceptable, for if a customer arrives to find the fast server (only) occupied, his or her expected delay is $1/\mu_2$ if he or she opts for the slow server and $2/\mu_1$ if he

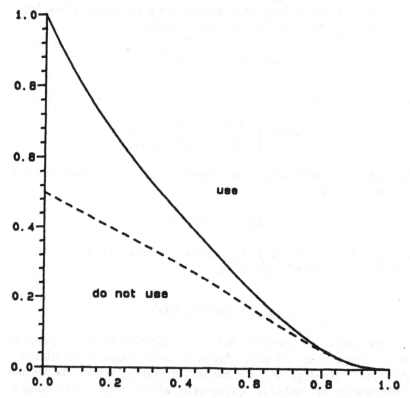

Figure 6.2. Should we use a slow server? The abscissa is λ/μ_1, the traffic intensity with just the fast server, and the ordinate is μ_2/μ_1, the ratio of the service rates. Above the line the mean delay time is less with two servers, below it is less with just the faster server. The solid line refers to indistinguishable servers, the dashed line refers to a preference for the fast server when both are free.

or she elected to wait for the fast server. (By the memoryless property of the exponential distribution, the remaining service time of the customer being served is still exponentially distributed with rate μ_1.)

Simulation can be used to study more general queueing systems. Let (A_i) be the interarrival times and (S_i) the service times in a FIFO single-server queue. We saw in Section 5.3 that the waiting time W_i of the ith customer satisfies

$$W_i = \max(0, W_{i-1} + S_{i-1} - A_i)$$

Then the delay time $D_i = W_i + S_i$, so

$$D_i = S_i + \max(0, D_{i-1} - A_i)$$

For a two-server queue we can use discrete-event simulation. The next event is either an arrival or a departure from one of the servers. By keeping track of the time of each of the three possible next events (possibly $+ \infty$ if a server is free), one can simulate the whole queueing system. (A little care is necessary to handle the event of a customer arriving when at least one server is free.) These methods work for a completely general arrival process and service-time distribution. At least 2000 customers could be simulated per CPU second on a VAX11/782.

We only considered a Poisson arrival process. Without any loss of generality one can take $\lambda = 1$ and measure time in minutes. Both queueing systems regenerate when a customer arrives to find all servers free, and regeneration is frequent at all but very heavy traffic rates. Figure 6.3 shows three runs with $\mu_1 = 1.25$, so the mean service time is 0.8 min. Runs are illustrated with a slow server with $\mu_2 = 0.625$ and $1/15$ as well as without the slow server. [In all cases the service time distribution was gamma (5), suitably scaled.] With $\mu_2 = 0.625$ there were 278 tours, whereas for $\mu_2 = 1/15$ there were only 32, the tours being lengthened by the long service times of the slow server. For exponential service times Rubinovitch gives the probability that a customer finds the queue empty, illustrated in Fig. 6.4. From this we can find the mean tour length as the reciprocal, using ergodicity.

In comparing queueing systems it is tempting to use a common arrival process, as was done in Fig. 6.3. This may complicate the analysis. For example, when considering whether to use a slow server we have a regeneration point only when customer i finds the queue empty in *both* systems. This reduces the frequency of regeneration points, possibly too much to allow a regenerative analysis.

Figure 6.5 shows autocorrelation plots for the three runs of Fig. 6.3. The long delay times with a slow server inflate the variance and reduce the autocorrelations. For autocorrelation-based analysis it is the single-server system that provides the larger problem, in contrast to regenerative simulation. It is rather difficult to use any of the methods of Section 6.1 to discard an initial transient. The disparity between the two servers can cause cyclic oscillations in the delay times from customer to customer. The problem was circumvented by starting the simulations in a "typical" state from a pilot run. An alternative would be to rely on analytical results from exponential service times that show rapid convergence to equilibrium.

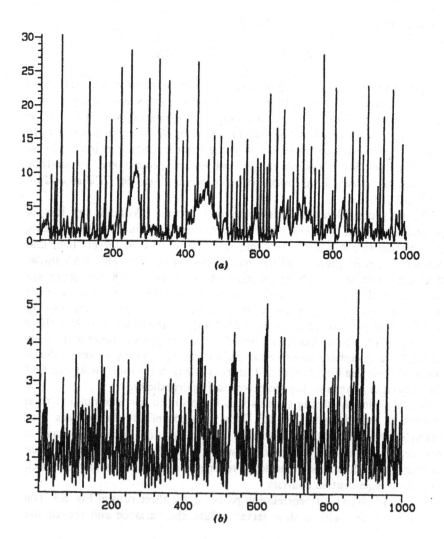

(a)

(b)

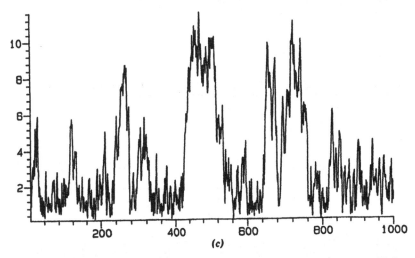

Figure 6.3. Plots of D_i vs i for 1000 customers with (a) two servers, $\mu_1 = 1.25$, $\mu_2 = 1/15$; (b) two servers, $\mu_1 = 1.25$, $\mu_2 = 0.625$; and (c) one server with $\mu = 1.25$. In all cases $\lambda = 1$ with a Poisson arrival process and gamma (5) service times.

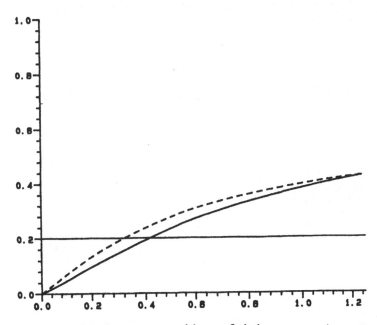

Figure 6.4. Probability of a customer arriving to find the queue empty versus μ_2 with $\mu_1 = 1.25$. The horizontal line is for just the fast server, the solid curve is for indistinguishable servers and the dashed curve is for a preference for the fast server.

165

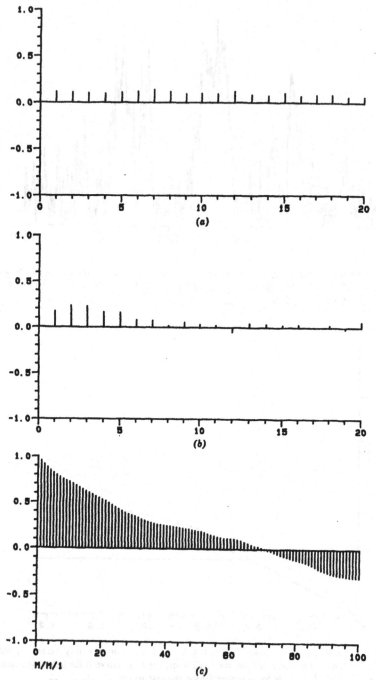

Figure 6.5. Autocorrelation plots for the data of Figure 6.3.

166

Tables 6.4 and 6.5 show some results based on Fig. 6.3. The two measures considered are mean delay time and the proportion of customers delayed by more than 5 min. The value $\mu_2 = 1/15$ was chosen to give virtually identical mean delay times with or without the slow server under exponential service times. In almost all cases the standard errors are seriously underestimated if the results are assumed to be independent. The exception is the long delays in Fig. 6.3b, which are very rare and so virtually independent. The agreement between the remaining methods of estimating the standard error is encouraging. These values are themselves not very accurate. For example, in the last column of Table 6.4 we have 95% confidence intervals of (0.37, 0.97) for the

Table 6.4.   Mean Delay Time and Estimates of its Standard Error for the Data of Fig. 6.3

	Fig. 6.3a	Fig. 6.3b	Fig. 6.3c
Mean	3.27	1.40	3.30
Standard deviation	3.99	0.835	2.75
Standard errors			
Under independence	0.126	0.026	0.087
Via 10 replications	0.264	0.038	0.52
Via $\tau^2 = 2\pi f(0)$	0.290	0.050	0.58
Via regeneration	0.257	0.049	0.64
Via regeneration, jacknifed	0.289	0.050	0.73

Table 6.5.   Estimates of θ = Probability that a Customer is Delayed more than 5 min, for the Data of Fig. 6.3

	Fig. 6.3a	Fig. 6.3b	Fig. 6.3c
θ	17%	0.2%	23.4%
s.e.(θ)			
Under independence	1.2%	0.141%	1.34%
Via 10 replications	4.5%	0.143%	7.6%
Via regeneration	3.5%	0.137%	9.7%
Via regeneration, jacknifed	3.9%	0.139%	11.0%

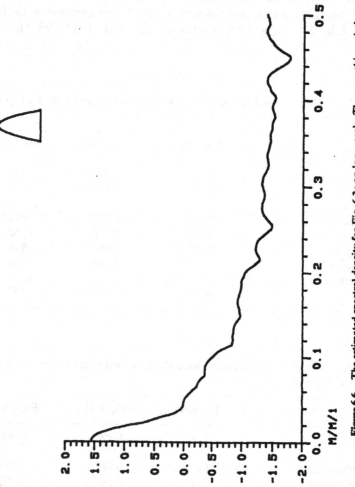

Figure 6.6. The estimated spectral density for Fig. 6.3c on $\log_{10}$ scale. The smoothing window and a pointwise 95% confidence interval are shown.

replications estimate and (0.49, 0.69) for the spectral density estimate (Fig. 6.6). Note that by using an internal estimate via spectral densities we do as well from one run as we would do from 70 replications.

These runs were picked from a larger study to illustrate some typical behavior. Although a run of 1000 customers is cheap, longer runs raise problems of storing the results and analyzing them by standard packages. As batches of length 1000 are certainly virtually independent, analyses of the type illustrated were performed on several batches and averaged. This gave acceptably accurate estimates of the quantities considered without excessive work in estimating standard errors. The only variance reduction that seemed worthwhile was to use a common arrival stream. For mean delay times one can analyze the differences in delay times for each customer. For the proportion of long waiting times no variance reduction was attempted, since the cause of long delays differs in the two systems, and any valid analysis of the combined system is rather complicated.

EXERCISES

6.1. Consider an $AR(1)$ process of 100 observations. Compute $\text{var}(\bar{Z})$, τ^2/n, and Es^2/n. How large need $|\alpha|$ be for τ^2/n not to be a good estimate of $\text{var}(\bar{Z})$?

6.2. Derive Moran's formulas for $E\hat{V}_1$ and $E\hat{V}_2$.

6.3. Simulate the process $X_t = 10 + \alpha(X_{t-1} - 10) + \varepsilon_t, \varepsilon_t \sim N(0,1)$ from $X_0 = 0$ and see how well you can choose how much to discard. [Note that $EX_t = 10(1 - \alpha^t)$.]

6.4. Suppose $\text{var}(Y_i|X_i) = \gamma X_i$ in the regenerative analysis. Show that $\text{var}(\hat{\theta}) \approx \gamma/\bar{X}$ and that γ can be estimated by the variance of $(Y_i - \hat{\theta}X_i)/\sqrt{X_i}$.

6.5. Estimate the constant c in Meketon and Heidelberger's formula for the data of Table 1.3. Hence adjust $\hat{\theta}$ for bias ($t = 200$ here) and compare with the jacknife estimate $\tilde{\theta}$.

6.6. If you have access to a discrete-event simulation language or system, check what assumptions it makes in its output analysis. Are they reasonable?

6.7. The best way to understand output analysis is to try it. Either simulate the processes of Section 6.5 or a problem from your own field and try several different output analyses.

CHAPTER 7

Uses of Simulation

In a sense this whole book is about uses of simulation. What concerns us here are some of the less obvious uses of simulation, which fall into two broad categories. The first category is within statistics, to perform statistical inference. Obvious uses of simulation in statistics are those of randomization such as randomizing experiments and randomized tests. More innovative uses are Monte-Carlo tests, the bootstrap, and Monte-Carlo confidence intervals. Often when a distribution is unknown, for example that of a test statistic, it is tempting to replace it by a distribution estimated from a simulated sample. Undoubtedly this has been done for many years in an *ad hoc* way. Increased computer power has made it possible on a large scale, and more formal methods such as the bootstrap and Monte-Carlo tests have been developed. These are the subject of Section 7.1.

Stochastic algorithms have recently proved successful in both cryptography and optimization. Although there is a certain appeal about an algorithm that will always succeed, in practice we may be able to afford to solve a much larger problem with a probabilistic algorithm that has only a high probability of success. Consider, as an example, the problem of finding whether a large integer is prime or the product of a small number of large primes. (Small prime factors can be found by conventional means.) There are stochastic algorithms that will report correctly if the number is prime, and find a factor if one exists with probability at least one-half. As these algorithms are very much cheaper than any that give a definite answer, we can afford to run the algorithm 20 times. We will then either know a factor or have odds over a million to one that the number is a prime (Devlin, 1984, pp. 176–178). This will usually suffice! Stochastic algorithms in optimization usually depend on the space-filling properties of random walks to reach a global rather than local optimum. This is considered further in Section 7.2.

Monte-Carlo integration was used as a test problem for variance reduction in Chapter 5. We were unable to escape the precision proportional to $1/\sqrt{n}$ law, although the constant could be made small. Deterministic methods can do much better in one or two dimensions, and we might give up

170

independence to try to do as well. Section 7.4 discusses quasi-Monte-Carlo integration in which the pseudo-random numbers are bent to fit the particular problem.

7.1. STATISTICAL INFERENCE

An appealing but potentially expensive way to develop statistics would be to compare the data with samples from the models under consideration. Since this would have to be done for each parameter value, some analytical aid is necessary. Yet ideas increasingly along that road are being developed.

Monte-Carlo Tests

Suppose we have a completely specified model and a goodness-of-fit statistic T for which small values indicate departures from the model. The random spatial pattern of Chapter 1 provides an example, with $T = n(n - 1)d^2$. To perform a pure significance test we need to know the distribution of T. This may be difficult or impossible analytically, but we can always simulate from the model and produce m samples $t_1, \ldots, t_m$ of T under the null hypothesis. One way to proceed would be to estimate the cdf of T by the empirical cdf of $(t_1, \ldots, t_m)$ or a smoothed version thereof. Essentially we estimate the critical point at level α as the 100αth percentile of $(t_1, \ldots, t_m)$.

Monte-Carlo tests are a closely related (but not identical) idea. If the null hypothesis is true we have $m + 1$ samples from the distribution of T, m by simulation and one by observation. Thus the probability that T is the kth smallest or smaller is $k/(m + 1)$ provided we can ignore ties. [For the rest of this section we assume a continuous distribution for T to avoid such difficulties; Jöckel (1986) shows how they can be resolved.] If we choose k and m to obtain a conventional significance level (say 1% or 5%), we have Barnard's (1963) Monte-Carlo test. [Dwass (1957) gave a special case earlier and the idea has been rediscovered since, but Barnard appears to have been the first to publish it explicitly.] Two-sided tests can be developed in exactly the same way.

The performance of Monte-Carlo tests has been considered by Hope (1968), Birnbaum (1974), Marriott (1979), and Jöckel (1984, 1986). The Monte-Carlo test has a random critical point and so "blurs" the critical region. Let F be the cdf of T under the null hypothesis. Then $U = F(T) \sim U(0, 1)$, and the conventional test rejects H_0 if $U < \alpha$. On the other hand, the Monte-Carlo test rejects H_0 with probability $p(U)$, where

$$p(u) = \sum_{r=0}^{k-1} \binom{m}{r} u^r (1 - u)^{m-r}$$

This should be interpreted as the proportion of times the Monte-Carlo test will reject with $T = F^-(u)$ as the observation. Marriott (1979) tabulates $p(u)$ and shows that for small k the blurring can be substantial. Note, however, that the Monte-Carlo and conventional tests only give different decisions a significant proportion of the time when T corresponds to a p-value near the significance level α.

An alternative approach is to consider the power function $\beta^m(\alpha)$ of the Monte-Carlo test versus the power $\beta(\alpha)$ of the conventional test. We would expect $\beta^m \leqslant \beta$ and find this to be the case except when the conventional test is worse than useless (Foutz, 1980, 1981; Jöckel, 1981). The important question is how large the loss of power can be. Since the Monte-Carlo test rejects if and only if

$$ t < t_{(k)} $$

the power under an alternative, F_θ, is

$$ \beta^m(\alpha) = \int_{-\infty}^{\infty} I(F^-(U_{(k)}) > t) dF_\theta(t) $$

$$ = \int_{-\infty}^{\infty} \int_{F(t)}^{1} b(\alpha, m, \xi) d\xi dF_\theta(t) $$

$$ = \int_{0}^{1} F_\theta(F^-(\xi)) b(\alpha, m, \xi) d\xi $$

where $b(\alpha, m, \cdot)$ is the pdf of the beta distribution with parameters $\alpha(m + 1)$ and $(1 - \alpha)(m + 1)$. Since $\beta(\alpha) = F_\theta(F^-(\alpha))$ we find

$$ \beta^m(\alpha) = \int_{0}^{1} \beta(\xi) b(\alpha, m, \xi) d\xi \tag{1} $$

Jöckel (1984, 1986) exploits this formula. For example,

Theorem 7.1. Suppose $\beta(\ )$ is concave on $[0, 1]$. Then $\beta^m(\alpha) \uparrow \beta(\alpha)$ as $m \to \infty$.

PROOF. From (1)

$$ \beta^{m+1}(\alpha) - \beta^m(\alpha) = \int_{0}^{1} \{b(\alpha, m + 1, \xi) - b(\alpha, m, \xi)\} \beta(\xi) d\xi $$

There are points ξ_1, ξ_2 with $0 < \xi_1 < \xi_2 < 1$ such that the integrand is zero at these points, positive on (ξ_1, ξ_2) and negative on $(0, \xi_1)$ and $(\xi_2, 1)$. Let

$$L(\xi) = \beta(\xi_1) + [\beta(\xi_2) - \beta(\xi_1)][\xi - \xi_1]/(\xi_2 - \xi_1)$$

so $L \leqslant \beta$ on (ξ_1, ξ_2) and $L \geqslant \beta$ elsewhere (by concavity). Then

$$\beta^{m+1}(\alpha) - \beta^m(\alpha) \geqslant \int_0^1 L(\xi)\{b(\alpha, m+1, \xi) - b(\alpha, m, \xi)\}d\xi = 0$$

as required, since both densities have mean α.   □

If $\beta(\ ^{-}\ )$ is continuous at α, the convergence of $\beta^m(\alpha)$ to $\beta(\alpha)$ follows directly from (1) (Hope, 1968; Birnbaum, 1974) by the L^2 convergence of $b(\alpha, m, \cdot)$ to α. How much do we lose by using a Monte Carlo test? Jöckel gives,

Theorem 7.2. Suppose $\beta(\ )$ is concave on $[0, 1]$, $\beta(0) = 0$ and $\beta(1) = 1$. Then

$$\frac{\beta^m(\alpha)}{\beta(\alpha)} \geqslant 1 - \frac{E|Z - \alpha|}{2\alpha} \approx 1 - \left[\frac{(1 - \alpha)}{2\pi m \alpha}\right]^{1/2}$$

where $Z \sim \text{beta}(\alpha(m + 1), (1 - \alpha)(m + 1))$.

PROOF.

$$\beta(\alpha) - \beta^m(\alpha) = E[\beta(\alpha) - \beta(Z)] \leqslant E[\beta(\alpha) - \tilde{\beta}(Z)]$$

where $\tilde{\beta}(\alpha)$ is the function linear on $[0, \alpha]$, $[\alpha, 1]$ agreeing with β at $\{0, \alpha, 1\}$. Then $\tilde{\beta}$ is the power of the randomized test

$$\phi_\xi = \begin{cases} \alpha^{-1}\xi\phi_\alpha, & \xi \leqslant \alpha \\ \dfrac{1 - \xi}{1 - \alpha}\phi_\alpha + \dfrac{\xi - \alpha}{1 - \alpha}, & \xi > \alpha \end{cases}$$

Then

$$E[\beta(\alpha) - \tilde{\beta}(Z)] = E\left[\frac{\beta(\alpha) - \alpha}{2\alpha(1 - \alpha)}|Z - \alpha| + \frac{2\alpha\beta(\alpha) - \beta(\alpha) - \alpha}{2\alpha(1 - \alpha)}(Z - \alpha)\right]$$

$$= E\left[\frac{\beta(\alpha) - \alpha}{2\alpha(1 - \alpha)}|Z - \alpha|\right] \leqslant \frac{\beta(\alpha)(1 - \alpha)}{2\alpha(1 - \alpha)}E|Z - \alpha|$$

since $EZ = \alpha$. The approximation comes from the central limit theorem, approximating Z by a normal variate of mean $\alpha(1 - \alpha)/m$.   □

Jöckel (1986) gives other more accurate formulas for the lower bound. For $\alpha = 5\%$ the bound goes from 64% at $m = 19$ through 83% at $m = 99$ to 94.5% at $m = 999$.

This formula gives a worst-case bound. For more specific results we have to consider asymptotic theory. Jöckel shows that the local asymptotic relative Pitman efficiency is

$$\left[\frac{\int_0^1 \phi(\Phi^{-1}(u))b(\alpha, m, u)du}{\phi(\Phi^{-1}(\alpha))} \right]^2$$

at least in the normal limit case. Again for $\alpha = 5\%$ the efficiency is 81% at $m = 19$ and 95.6% at $m = 99$. Further details are in Jöckel (1986).

These results generally confirm our heuristic ideas about Monte-Carlo tests. They provide some reasons to avoid small values of k, which blur the critical region rather a lot and so lose efficiency and power. Of course, since $(m + 1) = k/\alpha$, this implies m large. Unless one takes a very rigid approach to significance testing, $m = 99$ is usually sufficient.

In some cases one can make a decision without generating all m samples. In the spatial randomness example of Chapter 1 we commented that we could stop if we found *one* small distance. Analogously we can stop when k samples t_i are smaller than the observed T, for then we know we will not reject. Similarly but less usefully if $m - k + 1$ values exceed T, we will certainly reject and can stop. This device can reduce the labor needed if the fit is good, so T is a typical value, but will help little if the true p-value for T is small.

Efron's Bootstrap

Efron (1979) introduced an idea analogous to the jacknife that has aroused considerable interest as a general way to estimate a sampling distribution. Suppose we have an estimator $\hat{\theta}$ of θ based on a sample $x_1, \ldots, x_n$. The suggestion is to estimate the bias and variance of $\hat{\theta}$ by replacing the unknown distribution of $\hat{\theta}$ by the distribution of $\hat{\theta}$ under resampling from $(x_1, \ldots, x_n)$. That is, we draw a new sample $(y_1, \ldots, y_n)$ by sampling (with replacement) from $(x_1, \ldots, x_n)$ and compute $\hat{\theta}$ from the y_i's. This has a distribution from the random selection of the y_i's, and it is this distribution that is used in bootstrap methods. It *may* be possible to compute this resampling distribution analytically, but it is quick and easy to build up enough of a picture of it by simulation. The cost is in computer time, since for each resample we calculate a complicated estimator $\hat\theta$ and we will need to do this a large number m times. The simulation in contrast is trivial.

At first sight this is a preposterous idea, but it has been shown to work well in a wide variety of problems. Consider the very simple example of $\theta = \bar{x}$. We estimate var($\hat\theta$) by var($\bar{y}$), the variance under randomization. Each y_i is independently one of $\{x_1,\ldots,x_n\}$, so var(y_i) $= \sum(x_i - \bar{x})^2/n$ and

$$\text{var}(\bar{y}) = \sum(x_i - \bar{x})^2/n^2 = [(n - 1)/n]s^2/n$$

so this is (up to a factor close to 1) a sensible estimator. In general we will estimate var($\hat\theta$) by the sample variance of $\hat\theta(\mathbf{y})$ for a large number of samples. We can also estimate the bias of $\hat\theta$ or any other aspect of its distribution from the resampling distribution. For example, the bootstrap estimate of bias is

$$E\hat\theta - \theta$$

and the first term will be estimated by the mean of m resamples.

The regenerative simulation example of Section 6.4 provides a more realistic example. There are seven bivariate observations, $(1, 0)$, $(28, 4.692)$, $(12, 1.663)$, $(8, 0.387)$, $(26, 5.894)$, $(41, 15.922)$ and $(37, 22.787)$. There $\hat\theta$ is the ratio of the sum of the second component to the sum of the first component. Resampling 1000 times we find the bootstrap estimates of

$$\text{bias}(\hat\theta) = -0.011$$

$$\text{s.e.}(\hat\theta) = 0.087$$

The standard error agrees with previous estimates and the bias estimate is very close to that from the jacknife.

The bootstrap principle can only be justified asymptotically as $n \to \infty$ by showing that the resampling distribution and the true distribution of $\hat\theta$ have the same asymptotic behavior. The case of a finite sample space is easiest. We assume θ does not depend on the labeling of the sample. Let the possible values be $\{1,\ldots,L\}$. The distribution of the sample can be described by $\{f_j = P(x_1 = j)\}$ and the sample (x_i) can be described by $\{\hat{f}_j = (\text{no. of } x_i = j)/n\}$. Then the quantity of interest, $E\phi(\hat\theta) = EQ(\hat{f}, f)$, and this is estimated by $EQ(\hat{f}^*, \hat{f})$, where f^* refers to the resamples. Both $\hat{f}|f$ and $\hat{f}^*|\hat{f}$ have multinomial distributions, and as $n \to \infty$ both $\hat{f}^* - \hat{f}$ and $\hat{f} - f$ have the same asymptotic normal distribution. Under smoothness conditions on Q this gives the same asymptotic distribution for $Q(\hat{f}^*, \hat{f})$ as $Q(\hat{f}, f)$, as required. Later work has extended this result to more general problems.

Despite this asymptotic justification, bootstrap bias and variance estimates work well for small samples. [Efron (1982) gives many examples as evidence for this statement.]

Monte-Carlo Confidence Intervals

Monte-Carlo tests are only defined for a single simple null hypothesis and so cannot be inverted simply to form a confidence interval. Some pivotal quantity is needed. Suppose $\hat{\theta}$ is a consistent estimator of θ with cdf F_θ. Let θ^* be a sample from $F_{\hat{\theta}}$. We want to use the variation of θ^* about $\hat{\theta}$ to infer the variation of $\hat{\theta}$ about θ. Consider first a (local) location-family model

$$\hat{\theta} - \theta \sim F_0 \qquad \text{so } \theta^* - \hat{\theta} \sim F_0 \tag{2}$$

Let L and U be upper and lower $\frac{1}{2}\alpha$-prediction limits for θ^* obtained either analytically as

$$L = F_{\hat{\theta}}^-(\tfrac{1}{2}\alpha) = \hat{\theta} + F_0^-(\tfrac{1}{2}\alpha)$$
$$U = F_{\hat{\theta}}^-(1 - \tfrac{1}{2}\alpha) = \hat{\theta} + F_0^-(1 - \tfrac{1}{2}\alpha)$$

or via simulation from the empirical cdf of θ^*. The conventional $(1 - \alpha)$-confidence interval for θ is

$$\theta \in (\hat{\theta} - F_0^-(1 - \tfrac{1}{2}\alpha), \hat{\theta} - F_0^-(\tfrac{1}{2}\alpha))$$
$$= (2\hat{\theta} - U, 2\hat{\theta} - L) \tag{3}$$

if the model (2) holds exactly. If (2) is only a local approximation, (3) will be an approximate $(1 - \alpha)$ interval.

Suppose additionally that F_0 is symmetric about zero. Then analytically

$$U - \hat{\theta} = \hat{\theta} - L$$

and (3) becomes

$$\theta \in (L, U) \tag{4}$$

Efron (1982, Chapter 10) calls (4) a percentile confidence interval, and Buckland (1983, 1984) calls it a Monte-Carlo confidence interval.

An alternative assumption is a local scale family, particularly for $\theta, \hat{\theta} > 0$.

Then

$$\hat{\theta}/\theta \sim F_1, \qquad \theta^*/\hat{\theta} \sim F_1 \tag{5}$$

$$1 - \alpha = P_\theta(L/\hat{\theta} \leqslant \theta^*/\hat{\theta} \leqslant U/\hat{\theta})$$
$$= P_\theta(L/\hat{\theta} \leqslant \hat{\theta}/\theta \leqslant U/\hat{\theta})$$
$$= P_\theta(\hat{\theta}^2/U \leqslant \theta \leqslant \hat{\theta}^2/L)$$

giving the $(1 - \alpha)$-confidence interval

$$\theta \in (\hat{\theta}^2/U, \hat{\theta}^2/L) \tag{6}$$

If $\ln(\hat{\theta}/\theta)$ has a symmetric distribution we can again swap limits to obtain $\theta \in (L, U)$. Efron considers the assumption that there is a monotone increasing transformation g such that $g(\hat{\theta})$ has a symmetric location family of distributions with location parameter $g(\theta)$. This again gives $\theta \in (L, U)$. Suppose, however, that

$$g(\hat{\theta}) - g(\theta) \sim G$$

and G was symmetric about $m \neq 0$. Let L_g, U_g be percentiles of $g(\theta^*)$ as before, giving the confidence interval

$$g(\theta) \in (2g(\hat{\theta}) - U_g, 2g(\hat{\theta}) - L_g)$$

Using

$$U_g - g(\hat{\theta}) - m = -(L_g - g(\hat{\theta}) - m)$$

we obtain

$$g(\theta) \in (L_g - 2m, U_g - 2m)$$

Efron considers $g = \Phi^{-1} F^*_{\hat{\theta}}$, the asterisk denoting that this is estimated by the bootstrap. If we assume $g(\hat{\theta}) - g(\theta)$ is approximately a location family with point of symmetry $m = -\Phi^{-1} F^*_{\hat{\theta}}(\hat{\theta})$, we obtain

$$(F^{*-1}_{\hat{\theta}}\Phi(-2m - 2z_\alpha), F^{*-1}_{\hat{\theta}}\Phi(2z_\alpha - 2m)) \tag{7}$$

which is termed a "bias-corrected percentile interval."

Buckland (1984) claims the validity of (4) under the weaker condition

$$F_{\theta_1}(\theta_2) = 1 - F_{\theta_2}(\theta_1) \qquad \text{for all } \theta_1, \theta_2$$

Suppose $\psi = g(\theta)$ is a location family for an increasing transformation g, so $\hat{\psi} - \psi \sim G$. Then Buckland's condition becomes

$$P_G(\hat{\psi} \leqslant \psi_2 - \psi_1) = P_G(\hat{\psi} > \psi_1 - \psi_2)$$

that is, the symmetry of G about zero. Without such a pivotal relation we cannot conclude that $P_\theta(L \leqslant \theta \leqslant U) = 1 - \alpha$ for all θ from Buckland's condition.

Suppose that (4) is a valid confidence interval and that L and U are estimated as the percentiles from m simulations of θ^*. Then the achieved confidence level is

$$A = F_{\hat{\theta}}(U) - F_{\hat{\theta}}(L) = U_{(k)} - U_{(j)}$$

if the samples were generated by inversion from $(U_1, \ldots, U_m)$ and $j/(m + 1) \approx \frac{1}{2}\alpha$, $k/(m + 1) \approx 1 - \frac{1}{2}\alpha$. Then $EA = (k - j)/(m + 1) \approx 1 - \alpha$ and var $A \approx \alpha(1 - \alpha)/m$. (Note the divisor $m + 1$ rather than m.)

The intervals (3) and (6) appear to be new and remove many of the problems of their symmetrized cousin $\theta \in (L, U)$. Conversely, one has to choose an appropriate transformation, but for large samples with $\hat{\theta}$ near θ there should be little difference among (3), (4), and (6). Each assumes less than asymptotic normality.

7.2. STOCHASTIC METHODS IN OPTIMIZATION

Stochastic models arise in optimization in two ways. We may have to consider maximizing a quantity that is only measurable with error. One example is response-surface designs. There we wish to find a combination of controls maximizing the output of a plant, but we can only estimate the output from an experiment. Another example is maximizing a likelihood or minimizing a goodness-of-fit statistic when the appropriate distributions are so complex that they must be estimated by simulation. Such ideas are usually termed stochastic approximation or stochastic optimization.

The other use of randomness is to explore the set over which a function is to be maximized in an efficient manner. Stochastic methods generally do well in very complex optimization problems; when enough is known about the problem deterministic methods will be more efficient. The startling success of simulated annealing in combinatorial optimization (see below) suggests that stochastic methods are under-used at present.

Random Search Algorithms

We assume that a function f is given over a domain $D \subset \mathbb{R}^n$. The global maximum $x^* \in D$ is a point at which f attains its maximum, that is,

$$f(x^*) = \sup_{x \in D} f(x)$$

A local maximum $\hat{x}$ is a point such that $f(x) < f(\hat{x})$ for all $x \in D$ with $0 < \|x - \hat{x}\| \leqslant \delta$ for some choice of δ. Deterministic optimization methods will in general only find a local maximum. However, *if* D is convex and f is concave then a local maximum is the global maximum, so it is sufficient to ·find a local maximum. Without such additional knowledge global optimization methods need to cover the whole of D in some sense. One appealing way to do so is to use a random walk.

Suppose f is differentiable with gradient (column) vector g. Then two random search algorithms are:

Algorithm 7.1. Generate a random vector V_i uniformly distributed on the surface of the unit sphere $S_{n-1} \subset \mathbb{R}^n$ and move to

$$x_{i+1} = x_i + \alpha_i \{ [f(x_i + c_i V_i) - f(x_i - c_i V_i)]/2c_i \} V_i$$

Here α_i is a step length parameter, and the term $\{\cdots\}$ represents a finite difference approximation to $g^T V_i$.

Algorithm 7.2

1. Generate N random vectors V_{ik} uniformly on S_{n-1}.
2. Choose k to maximize $f(x_i + c_i V_{ik})$.
3. Move to

$$x_{i+1} = x_i + \alpha_i \{ [f(x_i + c_i V_{ik}) - f(x_i)]/c_i \} V_{ik}$$

Again, the term $\{\cdots\}$ is a finite-difference approximation to $g^T V_{ik}$. Both algorithms are related to gradient ascent, in which the next step is along $g(x_i)$. Note that for c_i small enough both algorithms will always ascend and so are likely to be trapped in a local maximum. Their attraction is that they will not exhibit the "zigzagging" of gradient ascent.

The most commonly used global optimization algorithm is undoubtedly

Algorithm 7.3

1. Select N starting points $x_i \in D$ according to some distribution over D.
2. Run a local optimization algorithm from x_i to reach $\hat{x}_i$.
3. Choose x^* as the $\hat{x}_i$ with largest $f(\hat{x}_i)$.

In essence, we find the local maximum nearest each starting point and choose the highest. Obviously if we have any idea of where the global maximum is we can reflect this in the choice of distribution over D.

Pincus (1968, 1970) uses a Monte-Carlo approach, based on

Theorem 7.3. Suppose f is continuous with a unique global maximum at x^*, and D is compact. Then

$$x^* = \lim_{\lambda \to \infty} \frac{\int_D x \exp[\lambda f(x)]dx}{\int_D \exp[\lambda f(x)]dx}$$

PROOF. Fix $\varepsilon > 0$. By compactness and continuity we can find $\delta > 0$ such that if $N_\varepsilon = \{x \mid |x - x^*| < \varepsilon\}$ then $f(x) < f(x^*) - \delta$ for all $x \in D \backslash N_\varepsilon$, and $\eta > 0$ such that $f(x) \geqslant f(x^*) - \delta/2$ for all $x \in N_\eta$. We will show

$$\lim_{\lambda \to \infty} \frac{\int_D |x - x^*| \exp \lambda f(x) dx}{\int_D \exp \lambda f(x) dx} = 0$$

Let I_n and I_d denote the numerator and denominator. Then

$$I_d \geqslant \int_{N_\eta} \exp \lambda f(x) dx \geqslant \text{vol}(N_\eta) \exp \lambda [f(x^*) - \delta/2]$$

Now

$$I_n = \int_{N_\varepsilon} |x - x^*| \exp \lambda f(x) dx + \int_{D \backslash N_\varepsilon} |x - x^*| \exp \lambda f(x) dx$$

$$\leqslant \varepsilon I_d + \sup_{x \in D} |x - x^*| \text{vol}(D) \exp \lambda f(x^*)$$

Finally, $I_n/I_d \leqslant \varepsilon + \text{const} \exp(-\lambda \delta/2)$ which suffices. $\square$

We can then apply the Markov chain sampling methods of Section 4.6 to the pdf proportional to

$$\exp[\lambda f(x)]$$

over D, to which they are ideally suited. Then x^* is estimated by the mean of X_n, the position of the Markov chain after n steps.

Simulated Annealing

This was introduced by Kirkpatrick et al. (1983) as a device to obtain improved solutions to combinatorial optimization problems, such as the wiring between integrated circuits on a printed circuit board and the n-city traveling salesman problem [see also Bonomi and Luttin (1984)]. These problems all have a cost U to be minimized by a choice of a large number of interrelated finite decisions. The traveling salesman problem is to visit n cities in any order without returning so as to travel the shortest possible distance. At each city he or she has to decide where to go next. Thus a path is a sequence of cities contained in $\{1, \ldots, n\}^n$ with not all paths allowed. The cost

$$U(\mathbf{x}) = \sum_{i=2}^{n} d(x_{i-1}, x_i)$$

The key step is to set up a probability distribution on all paths by

$$P_\lambda(\mathbf{x}) = \text{const} \exp[-\lambda U(\mathbf{x})] \tag{8}$$

with forbidden paths having infinite cost. As $\lambda \to \infty$ it is clear that P_λ will concentrate on the optimum path(s) $\mathbf{x}^*$. This is just the discrete version of Pincus' procedure, and we can again apply Markov chain methods.

There is a formal similarity between (8) and the pdf

$$P(\mathbf{x}) \propto \exp\left[-\frac{1}{kT} U(\mathbf{x})\right]$$

of the Gibbs measure in statistical physics of a configuration of energy U at temperature T. Annealing is a manufacturing process in which a molten metal is cooled extremely slowly to produce a stress-free solid. In mathematical terms this means achieving a configuration of atoms with energy near or at the minimum of U. This is precisely what we wish to do in the optimization problem and suggests taking $\lambda \propto 1/T$ and so increasing $\lambda \to \infty$ very slowly. If we are using the Markov chain method of sampling from (8) we will have to let this near equilibrium before λ is changed appreciably, and as $\lambda \to \infty$ the rate of convergence to equilibrium slows dramatically (just as in real annealing). Geman and Geman (1984) and Gidas (1985) study the choice of the sequence of T or, equivalently, of λ. Their methods are a detailed study of nonhomogeneous Markov chains and suggest in general that $\lambda \propto \log(1 + t)$, where t is the number of steps of the chain completed.

Unfortunately this is so slow a rate of convergence that there is no possibility of achieving a sample from a P_λ that is nearly concentrated on x^*. However, all is not lost, for we can monitor $U(X_n)$. The success of simulated annealing in combinatorial optimization has been to find paths x with $U(x)$ appreciably below any previously known feasible path. Although P_λ will not be concentrated on x^*, it will give increased mass to the near-optimum paths and so there is a good chance that even a rather casual exploration of all paths by random sampling from P_λ will come up with a good path. Why not then just choose a large λ and sample directly from this (as suggested by Pincus)? For large λ the successive points x_n move very slowly in the space of allowed paths, and starting with a small λ allows the process to search around rapidly for the correct local maximum. Although Gidas' results guarantee that the global maximum will be approached eventually, this could take a very long time. Thus it is helpful to resort to the random start procedure of Algorithm 7.3.

Image restoration and labeling problems give rise to Gibbs measures very naturally as posterior distributions. Maximizing the posterior distribution amounts to estimating by the posterior mode. The optimization problem involved can be enormous. Consider an image made up of an $M \times M$ array of pixels, for $M = 64$–512. Each pixel has one of k true colors but is observed with noise. Figure 7.1 shows a small example with $M = 64$ and $k = 2$. If we assume as prior distribution the Markov random field with specification

$$\frac{\ln P(x_{ij} = \text{black} \mid \text{other colors})}{\ln P(x_{ij} = \text{white} \mid \text{other colors})}$$

$$= \beta[\text{number of black neighbors} - \text{number of white neighbors}]$$

and additive Gaussian white noise of error variance κ, we find the posterior distribution

$$P(\mathbf{x}) \propto \exp[-U(\mathbf{x})]$$

where

$$U(\mathbf{x}) = \frac{1}{2\kappa} \sum_{\text{pixels}} (Z_{ij} - \mu_{x_{ij}})^2 - \beta \sum_{\substack{\text{neighbor} \\ \text{pairs}}} 1(\text{same color})$$

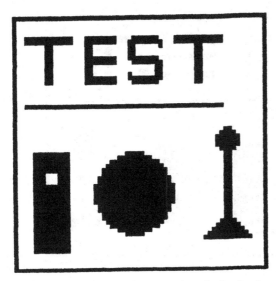

Figure 7.1.   A 64 × 64 test image for restoration via simulated annealing.

and Z_{ij} is the signal on pixel ij and μ_{white} and μ_{black} are the true luminance levels. This is easily simulated by the methods of Section 4.7. Figure 7.2 shows the results of 500 scans of the image with $\lambda = \ln(1 + \text{scan number})/3$. No local optimization method found a value of U within 100 of that for the image of Fig. 7.2.

Stochastic Optimization

So far we have considered stochastic algorithms for deterministic objective functions. Now suppose $f(x)$ can only be measured with error as $f(x, \varepsilon)$ and we are interested in solving

$$\max_{x \in D} E f(x, \varepsilon) \tag{9}$$

or

$$E f(x, \varepsilon) = 0 \tag{10}$$

Examples of this type of problem are given by Diggle and Gratton (1984) and Ruppert et al (1984). Both are concerned with parameter estimation in complex statistical models. For example, in maximum likelihood part of the

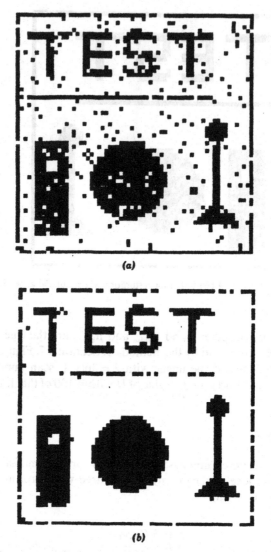

Figure 7.2. (a) Restoration with $\beta = 0$, that is, no spatial information was used. (b) "Minimum energy" restoration via simulated annealing with $\beta = 1$.

likelihood (such as a normalizing constant) may need to be estimated by simulation or in a moment-based estimator the theoretical moments could be found by simulation.

Some care is needed with the maximization problem. Suppose the noise is additive, $f(x, \varepsilon) = f(x) + \varepsilon$. Then any attempt to maximize $f(x, \varepsilon)$ directly will have to work with a rough function and will tend to overestimate $f(x^*)$, possibly appreciably. There are three main approaches, all of which make some use of the smoothness of $f(x)$.

The first of these approaches is response-surface methodology. An example is given by Hoel and Mitchell (1971). This was originally a way of designing a series of experiments to find the point of maximum yield of, say, a chemical plant (Box et al., 1978, Chapter 15). To solve (9) a local quadratic surface for $\phi(x) = Ef(x, \varepsilon)$ is assumed, with $E[f(x, \varepsilon) - \phi(x)]^2$ approximately constant. This quadratic can then be fitted by least squares and the design points arranged to close in on the predicted maximum. To solve (10) one would use a local linear approximation.

Perhaps the best-known approach is *stochastic approximation* (Wasan, 1969). The original form of Robbins and Monro (1951) solved (10) for $x \in R$ and ϕ increasing. A sequence (x_n) is used, defined recursively by

$$x_{n+1} = x_n - a_n f(x_n, \varepsilon_n)$$

Chung (1954) showed that (a_n) can be chosen so that $E(x_n - x^*)^2 = O(n^{-1})$, which is the best error rate achievable. Kiefer and Wolfowitz (1952) proposed solving (9) for concave ϕ by estimating $g(x, \varepsilon) = \partial f(x, \varepsilon)/\partial x$ by

$$g^*(x) = \tfrac{1}{2}[f(x + c_n, \varepsilon) - f(x - c_n, \varepsilon')]/c_n$$

and using the Robbins–Monro procedure for this function g^*. In general $Eg^*(x) \neq Eg(x, \varepsilon)$, so $c_n \to 0$ is necessary for $x_n \to x^*$. If ε_n and ε'_n are independent then the best achievable rate is $E(x_n - x^*)^2 = O(n^{-2/3})$ (Fabian, 1971). However, in a simulation study we will not need ε and ε' to be independent and by use of common random variables we can expect a high correlation between $f(x + c_n, \varepsilon)$ and $f(x - c_n, \varepsilon)$ leading to a much more stable estimator $g^*(x)$. This was exploited by Ruppert et al. (1984). Blum (1954) gave a multidimensional version of the Kiefer–Wolfowitz procedure.

Other authors, notably Springer (1969), Diggle and Gratton (1984), and Ruppert et al. (1984) generalize more sophisticated optimization methods, viewing the Kiefer–Wolfowitz–Blum procedure as a version of steepest ascent. Diggle and Gratton base their SQ procedure on the Nelder–Mead simplex algorithm whereas Ruppert et al. use a Newton-type procedure. (Note that Diggle and Gratton use a response-surface design near their optimum as a second phase.) Far too little experience is available to comment in any generality on a preferred procedure. These methods tend to be very expensive in computer time, so it is well worth tuning a method of stochastic optimization to the specific problem in hand. However, careful use of variance reduction along the lines suggested by Ruppert et al. has great potential benefits and may enable differences in $\phi(x) = Ef(x, \varepsilon)$ to be estimated sufficiently accurately to use derivative-based methods.

7.3. SYSTEMS OF LINEAR EQUATIONS

Monte-Carlo methods to solve the n by n system of equations

$$Ax = b \tag{11}$$

were proposed about 1950. They are generally inferior to the conventional methods of numerical analysis (which have been improved considerably since), but do have some value in specialized problems. Curtiss (1956) recommended the methods of this section only when a rough estimate of x was required or one was interested in only a few of the elements of x, for example in the diagonal elements of A^{-1}.

We can rewrite (11) as

$$x = Hx + b \tag{12}$$

for $H = I - A$ and consider solving (12) recursively by

$$x^{(k+1)} = Hx^{(k)} + b$$

from which we find

$$x^{(k+1)} = \sum_{r=0}^{k} H^r b + H^{k+1} x^{(0)} \tag{13}$$

This series will converge to x if $\rho(H) < 1$. (The spectral radius, ρ, is the largest magnitude of the eigenvalues where these exist.) We can ensure this by rescaling $A \to cA$ and $b \to cb$ and will assume from now on that (13) is convergent.

Monte-Carlo methods estimate H^r without forming the matrix product. The method used is analogous to importance sampling. Let P be the transition matrix of a Markov chain on $\{1,\ldots,n\}$ with $p_{ij} > 0$ if $H_{ij} \neq 0$. Suppose we sample $X_1,\ldots, X_k$ from this Markov chain with $X_0 = i$. Let

$$V_k = \prod_{j=1}^{k} \frac{h_{X_{j-1},X_j}}{p_{X_{j-1},X_j}} b_{X_k}$$

Then

$$E(V_k|X_0 = i) = \sum_{i_1} \cdots \sum_{i_k} h_{i,i_1} h_{i_1,i_2} \cdots h_{i_{k-1},i_k} b_{i_k} = (H^k b)_i$$

Thus if we take $x^{(0)}=0, X_0 = i$ and run the Markov chain for a long run k, we obtain an approximately unbiased estimator $\sum_0^k V_r$ of x_i. We can use the strong Markov property to obtain some information on x_j for each state j visited; start a product for x_j when $X_r = j$.

An alternative procedure is to look at the Markov chain in reverse. Choose a starting state X_0 from the distribution π on $\{1, \ldots, n\}$ with $\pi_i > 0$ if $b_i \neq 0$, and compute a product from $X_k, X_{k-1}, \ldots, X_0$. Let

$$W_k = \frac{b_{X_0}}{\pi_{X_0}} \prod_{j=1}^k \frac{h_{X_j, X_{j-1}}}{p_{X_{j-1}, X_j}}$$

Then, provided $p_{ij} > 0$ when $h_{ji} \neq 0$,

$$E(W_k \delta_{X_k i}) = \sum_{i_0} \cdots \sum_{i_k} b_{i_0} h_{i_1 i_0} \cdots h_{i_k i_{k-1}} \delta_{i_k i} = (H^k b)_i$$

For each run of the Markov chain we obtain approximately unbiased estimators of each element of x.

The original method of von Neumann and Ulam as published by Forsythe and Leibler (1950) differs slightly. Consider a Markov chain on $\{0, 1, \ldots, n\}$ with 0 an absorbing state and $1, \ldots, n$ transient. Then (X_n) reaches 0 almost surely. Again assume $p_{ij} > 0$ whenever $h_{ij} \neq 0$ and run the chain until it reaches state zero. Let X^* be the last state visited before reaching zero. Let

$$V^* = \prod_{j=1}^x \frac{h_{X_{j-1}, X_j}}{p_{X_{j-1}, X_j}} b_{X^*}$$

(adjoining ones to h to make a matrix on $\{0, 1, \ldots, n\}$). Then

$$E(V^* | X_0 = i) = \sum_{k=0}^x (H^k b)_i \frac{P(\text{absorb at step } k+1)}{p_{X_k, 0}} = x_i$$

Similarly if

$$W^* = \frac{b_{X_0}}{\pi_{X_0}} \prod_{j=1}^\infty \frac{h_{X_j, X_{j-1}}}{p_{X_{j-1}, X_j}}$$

then

$$E(W^* \delta_{X^* i}) = \sum_{k=0}^\infty (H^k b)_i \frac{P(\text{absorb at step } k+1)}{p_{X_k, 0}} = x_i$$

These give exactly unbiased estimators V^* and $W^* \delta_{x^*i}$. They are, however, more variable. (Note that $W^* \delta_{x^*i} \neq 0$ for only one value of i.) The idea of averaging over intermediate steps was due to Wasow (1952).

Halton (1962) considered accelerating the convergence of these schemes. His simplest and most effective idea was that having found a rough initial estimate $\tilde{x}$ of x one should solve

$$Ax = (b - A\tilde{x})$$

for a correction to $\tilde{x}$. This will have a much smaller right-hand side and so be solved more accurately. It is reminiscent of the ideas of iterative refinement in numerical analysis and of control variates.

All these schemes would be replicated to estimate element(s) of x. Rather little work has been done on the variances of the resulting estimators, which clearly depend on the choice of P. The variance need not even be finite and Halton (1970, pp. 17–23) discusses choosing P to make the variance finite.

One source of interest in Monte-Carlo methods of solving linear equations has been the field of integral equations, which when discretized give large systems of linear equations.

Eigenvalues

A related problem is to find eigenvalues of a $n \times n$ matrix A, that is, values λ such that

$$Ax = \lambda x$$

has a solution. Iterative methods of finding eigenvalues only find the most extreme eigenvalue; once this is found it is eliminated and the next most extreme found and so on. Suppose we start with an arbitrary vector $x^{(0)}$ and form

$$x^{(k+1)} = Ax^{(k)}$$

Suppose A is diagonalizable, so $A = C^T \Lambda C$ and

$$x^{(k)} = A^k x^{(0)} = C^T \Lambda^k C x^{(0)} \sim \lambda^k c^T x^{(0)} c$$

where λ is the extreme eigenvalue and c is the corresponding eigenvector. Thus we can estimate

$$|\lambda| \approx \| x^{(k+1)} \| / \| x^{(k)} \|$$

and guess the sign of λ by watching a few iterations. Again Monte-Carlo methods can be used to estimate $x^{(k)}$.

Another method is to renormalize $x^{(k+1)}$ by

$$\mu_{k+1} = \sum |x_i^{(k+1)}|, \qquad x^{(k+1)} \to x^{(k+1)}/\mu_{k+1}$$

Suppose $c \geqslant 0$. Then $\mu_{k+1} \to \lambda$, for $x^{(k)} \propto c$ approximately for large k. We can construct a stochastic version by taking a recurrent Markov chain on $\{1, \ldots, n\}$ with transition matrix Q satisfying $q_{ji} > 0$ if $a_{ij} \neq 0$. Let

$$W_k = \frac{x_{X_0}^{(0)}}{\pi_{X_0}} \prod_{j=1}^{k} \frac{a_{X_j, X_{j-1}}}{q_{X_{j-1}, X_j}}$$

Then

$$E(W_k \delta_{X_k, j} | X_0 = i) = (A^k)_{ji} x_i^{(0)}/\pi_i$$
$$E(W_k \delta_{X_k, j}) = (A^k x^{(0)})_j = \mu_1 \mu_2 \cdots \mu_k x_j^{(k)}$$
$$E(W_k) = \mu_1 \cdots \mu_k \sum x_j^{(k)} = \mu_1 \cdots \mu_k$$

for large enough k. Thus we can estimate λ by $(W_m/W_k)^{1/(m-k)}$ for $m > k$.

7.4. QUASI-MONTE-CARLO INTEGRATION

Pseudo-random numbers were defined in Chapter 2 by their pretence to randomness, and Monte-Carlo integration was defined in Section 5.1 assuming independent random samples, which gave an error of order $1/\sqrt{N}$ from N samples. Can we do better by giving up any pretence at mimicking randomness? The answer is that we can, by using quasi-random sequences and so performing quasi-Monte-Carlo integration.

Consider the one-dimensional integration problem

$$\theta = \int_0^1 f(x)dx$$

which we estimate by

$$\hat{\theta}_N = \frac{1}{N} \sum f(x_i)$$

for a sequence $x_1, \ldots, x_N$ from $[0, 1]$. We would like to choose the sequence to minimize $|\hat{\theta}_N - \theta|$ without needing excessive knowledge of f. For a set

$E \subset [0,1]$ let $v_N(E)$ be the proportion of the sequence within E. If f were the indicator function of E we would want $\Delta_N(E) = v_N(E) - \text{area}(E)$ to be small. In general, consider the following argument of Koksma (1942):

$$|\hat{\theta}_N - \theta| = \left| \int_0^1 f(x)dv_N(x) - \int_0^1 f(x)dx \right|$$

$$= \left| \int_0^1 f(x)d\Delta_N(x) \right|$$

$$= \left| \int_0^1 \Delta_N([0,x])df(x) \right|$$

after integration by parts. This gives two upper bounds. First

$$|\hat{\theta}_N - \theta| \leqslant \sup |\Delta_N([0,x])| \int d|f|(x)$$

the first term being the definition of D_N and the second the total variation of f. On the other hand, applying Cauchy–Schwartz,

$$|\hat{\theta}_N - \theta| \leqslant \left(\int_0^1 |\Delta_N([0,x])|^2 dx \right)^{1/2} \left(\int_0^1 |f'(x)|^2 dx \right)^{1/2}$$

$$\leqslant T_N \times \sup |f'(x)|$$

for a differentiable function f. Both D_N and T_N are known as *discrepancies*, with $T_N \leqslant D_N$.

For a random sequence (x_n) we have

$$ET_N^2 = \int_0^1 E|\Delta_N([0,x])|^2 dx$$

$$= \int_0^1 \frac{x(1-x)}{N} dx = 1/6N$$

So we would expect T_N to be of order $1/\sqrt{N}$. Let $F_N(x) = v_N([0,x])$ be the empirical distribution function of the random sequence. Then

$$D_N = \sup_x |F_N(x) - x|$$

is the Kolmogorov–Smirnov test statistic for a uniform distribution, which is well known to be of order $1/\sqrt{N}$. We can clearly do better by choosing a deterministic sequence. A little thought shows that the sequence of odd multiples of $1/2N$ is optimal and achieves $D_N = 1/2N$ and $T_N = 1/\sqrt{12N}$.

Discovering one should use a uniform grid in one dimension is no great advance, so we turn attention to $d \geqslant 2$ and the integral of f over $I = [0,1]^d$. There are several analogues of the bounds involving D_N and T_N. We can define $\Delta_N(x) = \Delta_N(\prod[0, x_i])$, $D_N = \sup|\Delta_N(x)|$, and $T_N = [\int_I \Delta_N(x)^2\, dx].^{1/2}$ The bounds involve the discrepancies of $(x_1,\ldots,x_n)$ and all $h < d$-dimensional subvectors extracted from (x_i). Halton (1970) and Niederreiter (1978) give full details. A weak form is

$$|\hat{\theta}_N - \theta| \leqslant D_N V(f)$$

where $V(f)$ is the total variation of f over I and all its subcubes.

This reduces the problem for rather general functions f to finding sets of points $(x_1,\ldots,x_N)$ with small D_N or T_N. A uniform grid in d dimensions has $D_N = O(N^{-1/d})$, $T_N = O(N^{-1/d})$ which for $d > 2$ will be worse than a random sequence. It is known that there are constants such that for any sequence

$$D_N \geqslant T_N \geqslant C_1^{(d)} N^{-1} (\ln N)^{(d-1)/2}$$

for all $d \geqslant 2$, and sequences are known with

$$D_N \leqslant C_2^{(d)} N^{-1} (\ln N)^{d-1}$$

and

$$T_N \leqslant C_3^{(d)} N^{-1} (\ln N)^{(d-1)/2}$$

For $d = 2$,

$$D_N \geqslant (N^{-1} \ln N)/(132 \ln 4)$$

These results [extracted from Niederreiter (1978)] show that with quasi-Monte-Carlo integration we can achieve

$$|\hat{\theta}_N - \theta| = O(N^{-1}(\ln N)^{d-1})$$

for smooth enough functions to satisfy an inequality for T_N, and

$$|\hat{\theta}_N - \theta| = O(N^{-1}(\ln N)^{d-1})$$

for functions of bounded total variation. Note that the sequences used depend on N, as in the one-dimensional case.

These quasi-random sequences seem to have been used only rarely, almost all reports being in the Russian literature. We can achieve better bounds if we are prepared to make stronger assumptions about f. Suppose we assume that f is periodic of period one in each variable x_i. Then we consider

$$\theta_N = \frac{1}{N} \sum_{r=1}^{N} f\left(\frac{r}{N} \mathbf{g}\right)$$

for a point $\mathbf{g} \in \mathbb{Z}^d$. By periodicity the summation points can be regarded as

$$\left(\frac{rg_1}{N} \bmod 1, \ldots, \frac{rg_d}{N} \bmod 1\right)$$

We recognize one such sequence with $\mathbf{g} = (1, a, a^2, \ldots, a^{d-1})^T$ from d-tuples of a congruential generator. Let c_h be the Fourier coefficients of f for $\mathbf{h} \in \mathbb{Z}^d$. Then

Theorem 7.4 (Korobov, 1959).   Let $r(\mathbf{h}) = \prod \max(1, |h_i|)$. Suppose

$$|c_h| \leqslant Cr(\mathbf{h})^{-k} \quad \text{for all } \mathbf{h} \neq 0 \tag{14}$$

Then

$$|\theta_N - \theta| \leqslant C P_k(\mathbf{g}, N)$$

where

$$P_k(\mathbf{g}, N) = \sum_{\{\mathbf{h} \neq 0, \mathbf{h}^T \mathbf{g} \equiv 0 \bmod N\}} r(\mathbf{h})^{-k}$$

It is known that there are vectors $\mathbf{g}$ such that $P_k(\mathbf{g}, N) = O(N^{-k}(\ln N)^{k(d-1)})$ but the proofs are nonconstructive. For $d = 2$ we consider $\mathbf{g} = (1, a)$. If $N = F_n$ for a Fibonacci number and $a = F_{n-1}$, we obtain $P_k(\mathbf{g}, N) = O(N^{-k} \ln N)$, a very much better order than more general methods. Specifically,

$$P_k(\mathbf{g}, F_n) \leqslant 120(\tfrac{8}{3})^{k-2} \frac{\ln F_n}{F_n^k} \quad \text{for } n \geqslant 5$$

a result of Zaremba. For $d \geqslant 3$ less is known but some tables of "good" $\mathbf{g}$ are available.

The conditions on Theorem 7.4 look restrictive. First, (14) holds if f is up to $(k-1)$ times jointly differentiable in each variable and all the mixed derivatives of these orders have bounded variation. Thus (14) is a smoothness condition. The periodicity can be removed by a transformation. For example,

$$f^*(x_1,\ldots,x_d) = 2^{-d} \sum_{\varepsilon_i = 0,1} f(\varepsilon_1 + (-1)^{\varepsilon_1} x_1,\ldots,\varepsilon_d + (-1)^{\varepsilon_d} x_d)$$

has the same integral as f but extends to a periodic function. Thus Theorem 7.4 provides a powerful multidimensional integration result. In contrast, a uniform grid of $N = n^d$ points achieves

$$|\hat{\theta}_N - \theta| = O(n^{-k}) = O(N^{-k/d})$$

a very much inferior error bound for large d.

The proofs of the results of this section depend on subtle number-theoretic arguments and are the province of a small band of specialists. Niederreiter (1978) gives an extensive bibliography and references for the results quoted here.

7.5. SHARPENING BUFFON'S NEEDLE

Buffon's needle experiment for the determination of π will be known to every reader. Although not a serious application of simulation, we will consider the long series of attempts to improve Buffon's experiment.

Buffon's original form was to drop a needle of length l "at random" on a grid of parallel lines of spacing d. For $l \leqslant d$

$$P(\text{needle intersects the grid}) = 2l/\pi d$$

To see this, let x be the distance from the center of the needle to the grid line below and θ be the angle of the needle to the horizontal. In any reasonable definition of "at random" $x \sim U(0, d), \theta \sim U(0, \pi)$ and they are independent. Thus

$$P = P(\text{needle intersects the grid})$$

$$= \pi^{-1} \int_0^\pi P(\text{needle intersects} \mid \theta = \phi) d\phi$$

$$= \pi^{-1} \int_0^\pi l \sin(\phi/d) d\phi$$

$$= 2l/\pi d$$

Let $\rho = l/d$, $\phi \doteq 1/\pi$. If we drop the needle n times and count R intersections,

$$\hat{p} = R/n, \qquad \text{var } \hat{p} = p(1-p)/n$$

$$\hat{\phi}_0 = \hat{p}/2\rho, \qquad \text{var } \hat{\phi}_0 = 2\rho\phi(1-2\rho\phi)/4\rho^2 n = \phi^2(1/2\rho\phi - 1)/n$$

so $\hat{\phi}_0$ is most accurate when $\rho = 1$. Thus if $\hat{\pi}_0 = 1/\hat{\phi}_0$,

$$\text{var } \hat{\pi}_0 \approx \pi^4 \text{ var } \hat{\phi}_0 \approx 5.63/n$$

Laplace considered replacing the grid of parallel lines by a grid of rectangles of sides a and b. A similar argument shows that (for $l \leqslant a, b$)

$$p_1 = P(\text{needle intersects the grid}) = \frac{2l(a+b) - l^2}{\pi ab}$$

Again var $\hat{\phi}_1$ is minimized by $a = b = l$, when $p_1 = 3/\pi$ and var $\hat{\phi}_1 = (1 - 3/\pi)/3\pi n$ and var $\hat{\pi}_1 \approx 0.47/n$.

An alternative suggested by Schuster (1974) is to count separately the number of intersections on the horizontal and vertical grid lines, with $l \leqslant a = b$. Let X_i and Y_i be the events that the ith drop intersects a horizontal and a vertical line respectively. Then Schuster proposed

$$\hat{p}_2 = \frac{1}{2n} \sum_1^n (X_i + Y_i)$$

as an unbiased estimator of $p_2 = 2\rho\phi$. Now $\text{cov}(X_i, Y_i) < 0$ since

$$P(X_i = 1) = P(Y_i = 1) = p$$

$$P(X_i = 1 \text{ and } Y_i = 1) = \rho^2/\pi = \rho^2 \phi$$

$$\text{cov}(X_i, Y_i) = \rho^2\phi - p^2 = \rho^2\phi - 4\rho^2\phi^2 = \rho^2\phi(1 - 4\phi)$$

We deduce that

$$\text{var}(\hat{p}_2) = \frac{1}{2n}[\text{var } X_i + \text{cov}(X_i, Y_i)]$$

$$= \frac{1}{2n}[p(1-p) + \rho^2\phi(1 - 4\phi)]$$

$$= \frac{1}{2n}[\rho\phi + \tfrac{1}{2}\rho^2\phi - 4\rho^2\phi^2]$$

so if $\hat{\phi}_2 = \hat{p}_2/2\rho$,

$$\text{var}(\hat{\phi}_2) = \frac{1}{n}\left[\frac{\phi}{4\rho} + \frac{\phi}{8} - \phi^2\right]$$

which is minimal for $\rho = 1$ as before. This gives

$$\text{var}(\hat{\pi}_2) \approx 1.76/n$$

We can also estimate ϕ by counting those needles that hit *both* sets of lines. This gives $\hat{\phi}_3 = \text{count}/\rho^2 n$ with

$$\text{var}(\hat{\phi}_3) = \rho^2 \phi(1 - \rho^2 \phi)/\rho^4 n$$

which is minimal for $\rho = 1$ and gives

$$\text{var}(\hat{\pi}_3) \approx 21.1/n$$

Could we have chosen *a priori* among $\hat{\phi}_0, \hat{\phi}_1, \hat{\phi}_2$, and $\hat{\phi}_3$? All are unbiased estimators and we could attempt to find a minimum variance unbiased estimator of ϕ. Perlman and Wichura (1975) show that in this problem the number of needles intersecting one or both sets of lines is a complete sufficient statistic for ϕ. Since this is equivalent to the number of needles hitting no lines, we are led to Laplace's formulation as the minimum variance unbiased estimator of ϕ. Since $\hat{\pi}_1$ is biased there still remains the possibility of a more efficient estimator of π in this set-up.

Replacing the parallel grid by a square grid is closely related to replacing the needle by a cross of side l (Hammersley and Morton, 1956). Let X and Y be indicator random variables of the events that each of the two needles of the cross intersects a grid line. Clearly for $l \leqslant d$ we find $EX = EY = 2\rho\varphi$ as before. As for the double grid we find $\text{cov}(X, Y) < 0$. The maximum precision of

$$\hat{\phi}_4 = \frac{1}{4\rho n}\sum(X_i + Y_i)$$

is attained for $l = d$ and leads to var $\hat{\pi}_4 \approx 2.42/n$. The corr$(X, Y) \approx -0.14$ is not large and we should consider the sufficient statistics in this problem. If the needles are indistinguishable there are three outcomes corresponding to

$Z = X + Y = 0, 1,$ or 2, with probabilities

$$P(Z = 0) = 1 - 2\sqrt{2}\phi$$
$$P(Z = 1) = 4(\sqrt{2} - 1)\phi$$
$$P(Z = 2) = 4(1 - 1/\sqrt{2})\phi$$

Let the number of occurrences of $Z = i$ be N_i in n drops. Then the likelihood is proportional to

$$(1 - 2\sqrt{2}\phi)^{N_0}((\sqrt{2} - 1)\phi)^{N_1}(4(1 - 1/\sqrt{2})\phi)^{N_2}$$

so again N_0 or $N_1 + N_2$ are sufficient statistics. This gives the estimator

$$\hat{p} = N_0/n, \qquad \hat{\phi}_s = \frac{1}{2\sqrt{2}}(n - N_0)$$

with

$$\text{var}(\hat{\phi}_s) = 0.0112/n$$

and

$$\text{var}(\hat{\pi}_s) \approx 1.09/n$$

which is better than antithetic variates but not as good as a single needle on a square grid for the same counting effort.

Yet another variation is to allow $l > d$ (Mantel, 1953). In this case we count the number of intersections of the grid by the needle. For a single grid the expected number of intersections is still $2l/\pi d$; divide the needle into parts of length less than d and add up. For a square grid the expected number is $4l/\pi d$. Let N be the number of intersections. Then

$$\text{var}(N) = \text{var}\{E[N|\theta]\} + E\{\text{var}(N|\theta)\}$$

The first term is $\text{var}\{\rho|\sin\theta|\} = \rho^2[\frac{1}{2} - 4/\pi^2]$ for a single grid. Conditional on θ the number of intersections is either $\text{int}(\rho|\sin\theta|)$ or $\text{int}(\rho|\sin\theta|) + 1$ so

$$\text{var}(N|\theta) = \text{frac}(\rho|\sin\theta|) - [\text{frac}(\rho|\sin\theta|)]^2 \leq \tfrac{1}{4}$$

and for large ρ $\mathrm{var}(N) \approx \rho^2(\frac{1}{2} - 4/\pi^2)$. This gives

$$\mathrm{var}(\hat{\pi}_6) \approx \frac{\pi^4}{4\rho^2 n} \mathrm{var}(N) \approx \frac{\pi^4}{4n}\left(\frac{1}{2} - \frac{4}{\pi^2}\right) \approx 2.31/n$$

for large ρ. In the case of a square grid we find

$$\mathrm{var}(\hat{\pi}_7) \approx \frac{\pi^4}{16\rho^2 n} \mathrm{var}\{E(N|\theta)\}$$

and $E(N|\theta) = \rho|\sin\theta| + \rho|\cos\theta|$ with variance $\rho^2(1 + 2/\pi - 16/\pi^2)$ so

$$\mathrm{var}(\hat{\pi}_7) \approx 0.094/n$$

Mantel suggested an alternative. Let s^2 be the sample variance of N. Then

$$Es^2 \approx \rho^2(1 + 2/\pi - 16/\pi^2)$$

for large ρ. Suppose we solve

$$s^2/\rho^2 = 1 + 2/\pi - 16/\pi^2$$

for π to obtain an estimator $\hat{\pi}$ with

$$\hat{\pi}_8 = (-1 + \sqrt{1 + 16c})/c$$

where $c = 1 - s^2/\rho^2$. Since $c \leqslant 1$, we obtain $\hat{\pi}_8 \geqslant 3.123$. Conversely, s^2 is bounded above by the case in which half the needles are parallel to the grid and half at $45°$ to the grid, giving $s^2/\rho^2 \leqslant (3 - 2\sqrt{2})/4$ and $\hat{\pi}_8 \leqslant 3.175$. We can estimate $\mathrm{var}(\hat{\pi}_8)$ using the large value of l to assume normality for the number of intersections, so s^2 is proportional to a χ^2_{n-1} variate. By the delta method we obtain

$$\mathrm{var}(\hat{\pi}_8) \approx (4 \times 10^{-3})/n$$

The estimators $\hat{\pi}_7$ and particularly $\hat{\pi}_8$ are the most accurate of those discussed. Against this we have only considered the limits for a large number of intersections on each needle, so the work for each drop is much greater than those of the earlier estimators. Because of the convergence at rate $1/\sqrt{n}$, Monte-Carlo methods are of course not a serious way to determine π but their history provides an interesting case of ingenuity in variance reduction.

EXERCISES

7.1. Formula (1) for the power of a Monte-Carlo test makes sense even when $k = \alpha/(m + 1)$ is noninteger. Show that this can be interpreted by a randomized Monte-Carlo test for which (1) holds.

7.2. Tabulate $p(u)$ for $u = 0.1\%$, 0.5%, 1%, 5%, 10%, and the common Monte-Carlo tests ($k = 1$, $m = 19$), ($k = 5$, $m = 99$), ($k = 1$, $m = 99$), ($k = 10$, $m = 199$), and ($k = 2$, $m = 199$).

7.3. Consider $H_0: T \sim N(0, 1)$ versus $H_1: T \sim N(\theta, I)$ for $\theta > 0$. For $\alpha = 5\%$ and $\alpha = 1\%$ compute the power of the Monte-Carlo test with $n = 99$ from (1) (by numerical or Monte Carlo integration) and plot against θ, together with the power of the conventional test and the lower bound given by Theorem 7.2. [See also Hope (1968).]

7.4. Many standard examples in statistical inference are either a location family or a scale family and so are not a good test of Monte-Carlo confidence intervals. One problem that is not is based on example (a) of Chapter 1. Consider n points in the unit square under an inhibition model that disallows distances less than R apart. Then asymptotically $n(n - 1)(d^2 - R^2) \sim \exp(\pi/2)$ where d is the minimum interpoint distance (Ripley and Silverman, 1978). Apply the Monte-Carlo confidence intervals (3), (4), (6), and (7) to R and compare with the conventional interval formed from R^2.

7.5. Apply Pincus' method to $f(x) = \sin x - x^2$ on $(-\pi, \pi)$. Try both selecting the next point of (X_n) uniformly on $(-\pi, \pi)$ (and so rejecting quite often) and uniformly on $(X_n - \delta, X_n + \delta)$ considered mod 2π, for small δ.

7.6. In the example after Theorem 2.14 we wished to invert

$$\frac{1}{512} \begin{bmatrix} 23 & 22 & 82 \\ 11 & -34 & 106 \\ -17 & 6 & 162 \end{bmatrix}$$

and knew that the inverse has integer elements. Use a Monte-Carlo method to estimate the inverse sufficiently accurately.

7.7.   Estimate the largest eigenvalue of

$$\begin{bmatrix} 12 & 6 & 6 \\ 11 & 5 & 8 \\ 1 & 9 & 14 \end{bmatrix}$$

given that it corresponds to a non-negative eigenvector.

7.8.   Show that $((2k - 1)/2N|k = 1,\dots, N)$ minimizes both D_N and T_N in one dimension.

7.9.   Use Zaremba's sequence to estimate

$$\theta = \int_{-1}^{1} \int_{-1}^{1} 1/(1 + x^2 + y^2)dy\,dx$$

7.10. In Laplace's version of Buffon's needle prove analytically or geometrically that p_1 is maximized by $a = b = l$.

7.11. Show that $N_1 + N_2$ is a complete sufficient statistic in the "single needle hitting a square $d \times d$ grid" problem even if $d \neq l$.

7.12. The drawback of Mantel's $\hat{\pi}_8$ is that it is consistent only as $l \to \infty$ as well as $n \to \infty$. Estimate the bias of $\hat{\pi}_8$ when $l = 10d$. In what direction will the bias be?

References

Ahrens, J. H. and Dieter, U. (1972) Computer methods for sampling from the exponential and normal distributions. *Comm. Assoc. Comput. Mach.* **15**, 873–882.

Ahrens, J. H. and Dieter, U. (1974) Computer methods for sampling from gamma, beta, Poisson and binomial distributions. *Computing* **12**, 223–246.

Ahrens, J. H. and Dieter, U. (1980) Sampling from binomial and Poisson distributions: a method with bounded computation times. *Computing* **25**, 193–208.

Ahrens, J. H. and Dieter, U. (1982a) Computer generation of Poisson deviates from modified normal distributions. *ACM Trans. Math. Soft.* **8**, 163–179.

Ahrens, J. H. and Dieter, U. (1982b) Generating gamma variates by a modified rejection technique. *Comm. Assoc. Comput. Mach.* **25**, 47–54.

Ahrens, J. H. and Kohrt, K. D. (1981) Computer methods for efficient sampling from largely arbitrary statistical distributions. *Computing* **26**, 19–31.

Andrews, D. F. (1976) Contribution to the discussion of Atkinson and Pearce (1976). *J. Roy. Statist. Soc.* A **139**, 452–453.

Andrews, D. F., Bickel, P. J., Hampel, F. R., Huber, P. J., Rogers, W. H., and Tukey, J. W. (1972) *Robust Estimates of Location.* Princeton University Press, Princeton.

Appleton, D. R. (1976) Contribution to the discussion of Atkinson and Pearce (1976). *J. Roy. Statist. Soc.* A **139**, 449–451.

Arnold, H. J., Bucher, B. D., Trotter, H. F., and Tukey, J. W. (1956) Monte Carlo techniques in a complex problem about normal samples. *Symposium on Monte Carlo Methods.* H. A. Meyer, Ed. Wiley, New York, pp. 80–88.

Arvillias, A. C. and Maritsas, D. G. (1978) Partitioning the period of a class of m-sequences and application to pseudorandom number generation. *J. Assoc. Comput. Mach.* **25**, 676–686.

Atkinson, A. C. (1977) An easily programmed algorithm for generating gamma random variables. *J. Roy. Statist. Soc.* A **140**, 232–234.

Atkinson, A. C. (1979a) The computer generation of Poisson random variables. *Appl. Statist.* **28**, 29–35.

Atkinson, A. C. (1979b) Recent developments in the computer generation of Poisson random variables. *Appl. Statist.* **28**, 260–263.

Atkinson, A. C. (1979c) A family of switching algorithms for the computer generation of beta random variables. *Biometrika* **66**, 141–146.

Atkinson, A. C. (1980) Tests of pseudo-random numbers. *Appl. Statist.* **29**, 164–171.

Atkinson, A. C. (1982) The simulation of generalized inverse Gaussian and hyperbolic random variables. *SIAM J. Sci. Stat. Comp.* **3**, 502–515.

Atkinson, A. C. and Pearce, M. C. (1976) The computer generation of beta, gamma and normal random variables (with discussion). *J. Roy. Statist. Soc.* A **139**, 431–461.

Atkinson, A. C. and Whittaker, J. (1976) A switching algorithm for the generation of beta random variables with at least one parameter less than 1. *J. Roy. Statist. Soc.* A **139**, 464–467.

Bailey, B. J. R. (1981) Alternatives to Hastings' approximation to the inverse of the normal cumulative distribution function. *Appl. Statist.* **30**, 275–276.

Baker, R. J. and Nelder, J. A. (1978) *GLIM Manual Release 3.* NAG, Oxford.

de Balbine, G. (1967) Note on random permutations. *Math. Comp.* **21**, 710–712.

Barker, A. A. (1965) Monte Carlo calculations of radial distribution functions for a proton-electron plasma. *Aust. J. Phys.* **18**, 119–133.

Barnard, G. (1963) Contribution to the discussion of Bartlett's paper. *J. Roy. Statist. Soc.* B **25**, 294.

Barr, D. R. and Slezak, N. L. (1972) A comparison of multivariate normal generators. *Comm. Assoc. Comput. Mach.* **15**, 1048–1049.

Bartels, R. (1978) Generating non-normal stable variates using limit theorem properties. *J. Stat. Comput. Siml.* **7**, 199–212.

Barton, D. E. and Mallows, C. L. (1965) Some aspects of the random sequence. *Ann. Math. Statist.* **36**, 236–260.

Bays, C. and Durham, S. D. (1976) Improving a poor random number generator. *ACM Trans. Math. Soft.* **2**, 59–64.

Beasley, J. D. and Springer, S. G. (1977) Algorithm AS111. The percentage points of the normal distribution. *Appl. Statist.* **26**, 118–121.

Bebbington, A. C. (1975) A simple method for drawing a sample without replacement. *Appl. Statist.* **24**, 136.

Bentley, J. L. and Saxe, J. B. (1980) Generating sorted lists of random numbers. *ACM Trans. Math. Soft.* **6**, 354–364.

Best, D. J. (1978) Letter to the editor. *Appl. Statist.* **27**, 181.

Best, D. J. (1979) Some easily programmed pseudo-random normal generators. *Aust. Comp. J.* **11**, 60–62.

Best, D. J. (1983) A note on gamma variate generators with shape parameter less than unity. *Computing* **30**, 185–188.

Beyer, W. A., Roof, R. B., and Williamson, D. (1971) The lattice structure of multiplicative congruential pseudo-random vectors. *Math. Comp.* **25**, 345–363.

Billingsley, P. (1968) *Convergence of Probability Measures.* Wiley, New York.

Birnbaum, Z. W. (1974) Computers and unconventional test statistics. *Reliability and Biometry: Statistical Analysis of Lifelength.* SIAM, Philadelphia, pp. 441–458.

Blomquist, N. (1970) On the transient behaviour of GI/G/1 waiting times. *Skand. Aktuar.,* 118–129.

Blum, J. R. (1954) Multidimensional stochastic approximation method. *Ann. Math. Statist.* **25** 737–744.

Bondesson, L. (1982) On simulation from infinitely divisible distributions. *Adv. Appl. Probab.* **14**, 855–869.

Bonomi, E. and Luttin, J.-L. (1984) The N-city travelling salesman problem: statistical mechanics and the Metropolis algorithm. *SIAM Review* **26**, 551–568.

Boswell, M. T. and DeAngelis, R. J. (1981) A rejection technique for the generation of random variables with the beta distribution. *Statistical Distributions in Scientific Work*, Vol. 4, C. Taillie et al., Eds. Reidel, Dordrecht, pp. 305–312.

Box, G. E. P. and Muller, M. E. (1958) A note on the generation of random normal deviates. *Ann. Math. Statist.* 29, 610–611.

Box, G. E. P., Hunter, W. G., and Hunter, J. S. (1978) *Statistics for Experimenters*. Wiley, New York.

Bratley, P., Fox, B. L., and Schrage, L. F. (1983) *A Guide to Simulation*. Springer-Verlag, New York.

Brent, R. P. (1974) A Gaussian pseudo random generator. *Comm. Assoc. Comput. Mach.* 17, 704–706.

Bright, H. S. and Enison, R. L. (1979) Quasi-random number sequences from a long-period TLP generator with remarks on application to cryptography. *Comp. Surv.* 11, 357–370.

Brillinger, D. R. (1973) Estimation of the mean of a stationary time series by sampling. *J. Appl. Probab.* 10, 419–431.

Brown, M. and Solomon, H. (1979) On combining pseudo-random number generators. *Ann. Statist.* 7, 691–695.

Buckland, S. T. (1983) Monte Carlo methods for confidence interval estimation using the bootstrap technique. *BIAS* 10, 194–212.

Buckland, S. T. (1984) Monte Carlo confidence intervals. *Biometrics* 40, 811–817.

Burt, J. M. Jr and Garman, M. B. (1971) Conditional Monte Carlo: a simulation technique for stochastic network analysis. *Manag. Sci.* 18A, 207–217.

Carson, J. S. and Law, A. M. (1980) Conservation equations and variance reduction in queueing simulations. *Oper. Res.* 28, 535–546.

Carter, G. and Ignall, E. J. (1975) Virtual measures: a variance reduction technique for simulation. *Manag. Sci. (Applications)* 21, 607–616.

Cassels, J. W. S. (1959) *An Introduction to the Geometry of Numbers*. Springer-Verlag, Berlin.

Cassels, J. W. S. (1978) *Rational Quadratic Forms*. Academic Press, London.

Chaitin, G. J. (1966) On the length of programs for computing finite binary sequences. *J. Assoc. Comput. Mach.* 13, 547–569.

Chambers, J. M. (1970) Computers in statistical research: simulation and computer-aided mathematics. *Technometrics* 12,, 1–15.

Chambers, J. M. (1977) *Computational Methods for Data Analysis*. Wiley, New York.

Chambers, J. M., Mallows, C. L., and Stuck, B. W. (1976) A method for simulating stable random variables. *J. Amer. Stat. Assoc.* 71, 340–344.

Chay, S. C., Fardo, R. D., and Mazumdar, M. (1975) On using the Box–Muller transformation with congruential pseudo-random number generators. *Appl. Statist.* 24, 132–135.

Chen, N. and Asau, Y. (1974) On generating random variates from an empirical distribution. *Amer. Inst. Ind. Eng. Trans.* 6, 163–166.

Cheng, R. C. H. (1977) The generation of gamma variables with non-integral shape parameters. *Appl. Statist.* 26, 71–74.

Cheng, R. C. H. (1978a) Generating beta variables with non-integral shape parameters. *Comm. Assoc. Comput. Mach.* 21, 317–322.

Cheng, R. C. H. (1978b) Analysis of simulation experiments under normality assumptions. *J. Oper. Res. Soc.* 29, 493–497.

Cheng, R. C. H. (1982) The use of antithetic variates in computer simulations. *J. Oper. Res. Soc.* **33**, 229–237.

Cheng, R. C. H. and Feast, G. M. (1979) Some simple gamma variate generators. *Appl. Statist.* **28**, 290–295.

Cheng, R. C. H. and Feast, G. M. (1980a) Gamma variate generators with an increased shape parameter range. *Comm. Assoc. Comput. Mach.* **23**, 389–394.

Cheng, R. C. H. and Feast, G. M. (1980b) Control variables with known mean and variance. *J. Oper. Res. Soc.* **31**, 51–56.

Chung, K. L. (1954) On a stochastic approximation method. *Ann. Math. Statist.* **25**, 463–483.

Church, A. (1940) On the concept of a random sequence. *Bull. Amer. Math. Soc.* **46**, 130–135.

Cox, D. R. and Isham, V. (1980) *Point Processes*. Chapman and Hall, London.

Cox, D. R. and Smith, W. L. (1961) *Queues*. Methuen, London.

Coveyou, R. R. and Macpherson, R. D. (1967) Fourier analysis of uniform random number generators. *J. Assoc. Comput. Mach.* **14**, 100–119.

Crane, M. A. and Iglehart, D. L. (1974a) Simulating stable stochastic systems I: general multi-server queues. *J. Assoc. Comput. Mach.* **21**, , 103–113.

Crane, M. A. and Iglehart, D. L. (1974b) Simulating stable stochastic systems II: Markov chains. *J. Assoc. Comput. Mach.* **21**, 114–123.

Crane, M. A. and Iglehart, D. L. (1975a) Simulating stable stochastic systems III: regenerative processes and discrete event simulation. *Oper. Res.* **23**, 33–45.

Crane, M. A. and Iglehart, D. L. (1975b) Simulating stable stochastic systems IV: approximation techniques. *Manag. Sci. (Theory)* **21**, 1215–1224.

Crane, M. A. and Lemoine, A. J. (1977) *An Introduction to the Regenerative Method for Simulation Analysis*. Lect. Notes Control Inform. Sci. **4**.

Cross, G. R. and Jain, A. K. (1983) Markov random field texture models. *IEEE Trans.* **PAMI-5**, 25–39.

Curtiss, J. H. (1956) A theoretical comparison of the efficiencies of two classical methods and a Monte Carlo method for computing one component of the solution of a set of linear algebraic equations. *Symposium on Monte Carlo Methods*. H. A. Meyer, Ed. Wiley, New York, pp. 191–223.

Davis, B. M., Hagan, R., and Borgman, L. E. (1981) A program for the finite Fourier transform simulation of realizations from a one-dimensional random function with known covariance. *Comp. Geosciences* **7**, 199–206.

Déak, I. (1980) Fast procedures for generating stationary normal vectors. *J. Stat. Comp. Siml.* **10**, 225–242.

Déak, I. (1981) An economical method for random number generation and a normal generator. *Computing* **27**, 113–121.

Devlin, K. (1984) *Microchip Mathematics*. Shiva, Nantwich.

Devroye, L. (1981) The computer generation of Poisson random variables. *Computing* **26**, 197–207.

Devroye, L. (1984) Random variate generation for unimodal and monotone densities. *Computing* **32**, 43–68.

Devroye, L. (1985) The analysis of some algorithms for generating random variates with a given hazard rate. *Nav. Res. Log. Q.*

Dieter, U. (1975) How to calculate shortest vectors in a lattice. *Math. Comp.* **29**, 827–833.

Dieter, U. and Ahrens, J. H. (1973) A combinatorial method for the generation of normally distributed random numbers. *Computing* 11, 137–146.

Diggle, P. J. and Gratton, R. J. (1984) Monte Carlo methods of inference for implicit statistical models (with discussion). *J. Roy. Statist. Soc.* B 46, 193–227.

Dixon, W. J. and Tukey, J. W. (1968) Approximate behavior of Winsorized t (Trimming/Winsorization 2). *Technometrics* 10, 83–98.

Dudewicz, E. J. and Ralley, T. G. (1981) *The Handbook of Random Number Generation and Testing with TESTRAND Computer Code*. American Sciences Press, Columbus, OH.

Duket, S. D. and Pritsker, A. A. B. (1978) Examination of simulation output using spectral methods. *Math. Comput. Simul.* 20, 53–60.

Durstenfeld, R. (1964) Algorithm 235. Random permutation. *Comm. Assoc. Comput. Mach.* 7, 420.

Dwass, M. (1957) Modified randomization tests for non-parametric hypotheses. *Ann. Math. Statist.* 28, 181–187.

Efron, B. (1979) Bootstrap methods: another look at the jacknife. *Ann. Statist.* 7, 1–26.

Efron, B. (1982) *The Jacknife, the Bootstrap and Other Resampling Plans*. SIAM, Philadelphia.

van Es, A. J., Gill, R. D., and van Putten, C. (1983) Random number generation for a pocket calculator. *Stat. Neerl.* 37, 95–102.

Fabian, V. (1971) Stochastic approximation. *Optimizing Methods in Statistics*. J. S. Rustagi, Ed. Academic Press, New York, pp. 439–470.

Fan, C. T., Muller, M. E., and Rezucha, I. (1962) Development of sampling plans using sequential (item by item) selection techniques and digital computers. *J. Amer. Statist. Assoc.* 57, 387–402.

Fellen, B. M. (1969) An implementation of the Tausworthe generator. *Comm. Assoc. Comput. Mach.* 12, 413.

Fermat, P. (1640) Letter to B. Frénicle. *Oeuvres* 2, 206–212.

Fishman, G. S. (1971) Estimating sample size in computer simulation experiments. *Manag. Sci.* 18A, 21–38.

Fishman, G. S. (1973) Statistical analysis for queueing simulations. *Manag. Sci. (Theory)* 20, 363–369.

Fishman, G. S. (1974) Estimation in multiserver queueing simulations. *Oper. Res.* 22, 72–78.

Fishman, G. S. (1978a) *Principles of Discrete Event Simulation*. Wiley, New York.

Fishman, G. S. (1978b) Grouping observations in digital simulation. *Manag. Sci.* 24, 510–521.

Fishman, G. S. (1983a) Accelerated accuracy in the simulation of Markov chains. *Oper. Res.* 31, 466–487.

Fishman, G. S. (1983b) Accelerated convergence in the simulation of countably infinite state Markov chains. *Oper. Res.* 31, 1074–1089.

Flinn, P. A. (1974) Monte Carlo calculation of phase separation in a 2-dimensional Ising system. *J. Statist. Phys.* 10, 89–97.

Forsythe, G. E. (1972) Von Neumann's comparison method for random sampling from normal and other distributions. *Math. Comp.* 26, 817–826.

Forsythe, G. E. and Leibler, R. A. (1950) Matrix inversion by the Monte Carlo method. *Math. Comp.* 4, 127–129; 5, 55.

Foutz, R. V. (1980) A method for constructing exact tests from test statistics that have unknown distributions. *J. Stat. Comp. Siml.* **10**, 187–193.

Foutz, R. V. (1981) On the superiority of Monte Carlo tests. *J. Stat. Comp. Siml.* **12**, 135–137.

Fréchet, M. (1951) Sur les tableaux de corrélation dont les marges sont données. *Ann. Univ. Lyon. sect.* A **14**, 53–77.

Freedman, D. (1971) *Markov Chains.* Holden-Day, San Francisco, CA.

Fuller, A. T. (1976) The period of pseudo-random numbers generated by Lehmer's congruential method. *Computer J.* **19**, 173–177.

Fushimi, M. and Tezuka, S. (1983) The *k*-distribution of generalized feedback shift register pseudorandom numbers. *Comm. Assoc. Comput. Mach.* **26**, 516–523.

Gardner, G., Harvey, A. C. and Phillips, G. D. A. (1979) Algorithm AS154. An algorithm for the exact maximum likelihood estimation of autoregressive-moving average models by means of Kalman filtering. *Appl. Statist.* **29**, 311–322.

Garfarian, A. V., Ancker, C. J., Jr, and Morisaka, T. (1978) Evaluation of commonly used rules for detecting "steady state" in computer simulation. *Nav. Res. Log. Q.* **24**, 667–678.

Garman, M. B. (1972) More on conditioned sampling in the simulation of stochastic networks. *Manag. Sci.* **19A**, 90–95.

Geman, S. and Geman, D. (1984) Stochastic relaxation, Gibbs distributions and the Bayesian restoration of images. *IEEE Trans.* PAMI-6, 721–741.

George, L. L. (1977) Variance reduction for a replacement process. *Simulation* **29**, 65–74.

Gerontidis, I. and Smith, R. L. (1982) Monte Carlo generation of order statistics from general distributions. *Appl. Statist.* **31**, 238–243.

Gidas, B. (1985) Nonstationary Markov chains and the convergence of the annealing algorithm. *J. Statist. Phys.* **39**, 73–131.

Gleser, L. J. (1976) A canonical representation for the noncentral Wishart distribution useful for simulation. *J. Amer. Statist. Assoc.* **71**, 690–695.

Golder, E. R. and Settle, J. G. (1976) The Box–Muller method for generating pseudo-random normal deviates. *Appl. Statist.* **25**, 12–20.

Golomb, S. W. (1967) *Shift Register Sequences.* Holden-Day, San Francisco.

Gonzalez, T., Sahni, S., and Franta, W. R. (1977) An efficient algorithm for the Kolmogorov-Smirnov and Lillefor tests. *ACM Trans. Math. Soft.* **3**, 60–64.

Good, I. J. (1953) The serial test for sampling numbers and other tests for randomness. *Proc. Camb. Phil. Soc.* **49**, 276–284.

Good, I. J. (1957) On the serial test for random sequences. *Ann. Math. Statist.* **28**, 262–264.

Greenwood, J. A. (1981) Algorithm 81-02. A portable formulation of the alias method for random numbers with discrete distributions. *Comm. Statist.-Simula. Comput.* B **10**(6), 649–655.

Greenwood, R. E. (1955) Coupon collector's test for random digits. *Math. Comp.* **9**, 1–5.

Gross, A. M. (1973) A Monte Carlo swindle for estimators of location. *Appl. Statist.* **22**, 347–353.

Gunter, F. L. and Wolff, R. W. (1980) The almost regenerative method for stochastic systems simulation. *Oper. Res.* **28**, 375–386.

Gustavson, F. G. and Liniger, W. (1970) A fast random number generator with good statistical properties. *Computing* **6**, 221–226.

Hacking, I. (1965) *Logic of Statistical Inference.* Cambridge University Press, London.

Halton, J. H. (1962) Sequential Monte Carlo. *Proc. Camb. Phil. Soc.* **58,**, 57–58.

Halton, J. H. (1970) A retrospective and prospective study of the Monte Carlo method. *SIAM Review* 12, 1–63.

Hammersley, J. M. and Handscomb, D. C. (1964) *Monte Carlo Methods*. Methuen, London.

Hammersley, J. M. and Morton, K. W. (1956) A new Monte Carlo technique: antithetic variates. *Proc. Camb. Phil. Soc.* 52, 449–475.

Hannan, E. J. (1957) The variance of the mean of a stationary process. *J. Roy. Statist. Soc.* B 19, 282–285.

Hastings, W. K. (1970) Monte-Carlo sampling methods using Markov chains and their applications. *Biometrika* 57, 97–109.

Heidelberger, P. and Iglehart, D. L. (1979) Comparing stochastic systems using regenerative simulation with common random numbers. *Adv. Appl. Probab.* 11, 804–819.

Heidelberger, P. and Welch, P. D. (1981a) A spectral method for confidence interval generation and run length control in simulations. *Comm. Assoc. Comput. Mach.* 24, 233–245.

Heidelberger, P. and Welch, P. D. (1981b) Adaptive spectral methods for simulation output analysis. *IBM J. Res. Dev.* 25, 860–876.

Heidelberger, P. and Welch, P. D. (1983) Simulation run length control in the presence of an initial transient. *Oper. Res.* 31, 1109–1144.

Heikes, R. G., Montgomery, D. C., and Rardin, R. L. (1976) Using common random numbers in simulation experiments—an approach to the statistical analysis. *Simulation* 27, 81–85.

Henery, R. J. (1983) Personal communication.

Hoaglin, D. C. and Andrews, D. F. (1975) The reporting of computation-based results in statistics. *Amer. Statist.* 29, 122–126.

Hoeffding, W. (1940) Masstabinvariante Korrelationstheorie. *Schriften des Mathematischen Instituts und des Instituts für Angewandte Mathematik der Universität Berlin* 5, 179–233.

Hoel, D. G. and Mitchell, T. J. (1971) The simulation, fitting and testing of a stochastic cellular proliferation model. *Biometrics* 27, 191–199.

Hope, A. C. A. (1968) A simplified Monte Carlo significance test procedure. *J. Roy. Statist. Soc.* B 30, 582–598.

Hopkins, T. R. (1983) Algorithm AS193. A revised algorithm for the spectral test. *Appl. Statist.* 32, 328–335.

Hordijk, A., Iglehart, D. L., and Schassberger, R. (1976) Discrete time methods for simulating continuous time Markov chains. *Adv. Appl. Probab.* 8, 772–778.

Hsuan, F. (1979) Generating uniform polygonal random pairs. *Appl. Statist.* 28, 170–172.

Hurst, R. L. and Knop, R. E. (1972) Algorithm 425. Generation of random correlated normal variables. *Comm. Assoc. Comput. Mach.* 15, 355–357.

Iglehart, D. L. (1975) Simulating stable stochastic systems V: comparison of ratio estimators. *Nav. Log. Res. Q.* 22, 553–565.

Iglehart, D. L. (1976) Simulating stable stochastic systems VI: quantile estimation. *J. Assoc. Comput. Mach.* 23, 347–360.

Iglehart, D. L. (1977) Simulating stable stochastic systems VIII: selecting the best system. *Algorithmic Methods in Probability* M. F. Neuts, Ed. North Holland, Amsterdam, pp. 37–50.

Iglehart, D. L. and Lewis, P. A. W. (1979) Regenerative simulation with internal controls. *J. Assoc. Comput. Mach.* 26, 271–282.

Iglehart, D. L. and Shedler, G. S. (1978) Regenerative simulation of response times in networks of queues. *J. Assoc. Comput. Mach.* **25**, 449–460.

Iglehart, D. L. and Shedler, G. S. (1980) *Regenerative Simulation of Response Times in Networks of Queues.* Lect. Notes Control Inform. Sci. **26**.

Iglehart, D. L. and Shedler, G. S. (1981) Regenerative simulation of response times in networks of queues: statistical efficiency. *Acta Informatica* **15**, 347–363.

Iglehart, D. L. and Shedler, G. S. (1983a) Statistical efficiency of regenerative simulation methods for networks of queues. *Adv. Appl. Probab.* **15**, 183–197.

Iglehart, D. L. and Shedler, G. S. (1983b) Simulation of non-Markovian systems. *IBM J. Res. Dev.* **27**, 472–480.

Iglehart, D. L. and Stone, M. L. (1983) Regenerative simulation for estimating extreme values. *Oper. Res.* **31**, 1145–1166.

Inoue, H., Kumahora, H., Yoshizawa, Y., Ichimura, M., and Miyitake, D. (1983) Random numbers generated by a physical device. *Appl. Statist.* **32**, 115–120.

Jackson, M. A. (1975) *Principles of Program Design.* Academic Press, London.

Jöckel, K.-H. (1981) A comment on the construction of exact tests from test statistics that have unknown null distributions. *J. Stat. Comp. Siml.* **12**, 133–134.

Jöckel, K.-H. (1984) Computational aspects of Monte Carlo tests. *Compstat 1984 Proc.*, Physica-Verlag, Vienna, pp. 183–188.

Jöckel, K.-H. (1986) Finite sample properties and asymptotic efficiency of Monte Carlo tests. *Ann. Statist.* **14**, 336–347.

Jöhnk, M. D. (1964) Erzeugung von Betaverteilen und Gammaverteilung Zufallszahlen. *Metrika* **8**, 5–15.

Johnson, D. E. and Hegemann, V. (1974) Procedure to generate random matrices with noncentral distributions. *Comm. Statist.* **3**, 691–699.

Jones, T. G. (1962) A note on sampling a tape-file. *Comm. Assoc. Comput. Mach.* **5**, 343.

Kabak, I. W. (1968) Stopping rules for queueing systems. *Oper. Res.* **16**, 431–437.

Kaminsky, F. C. and Rumpf, D. L. (1977) Simulating nonstationary Poisson processes: a comparison of alternatives including the correct approach. *Simulation* **29**, 17–20.

Kelly, F. P. (1979) *Reversibility and Stochastic Networks.* Wiley, Chichester, England.

Kemp, A. W. (1981) Frugal methods of generating bivariate discrete random variables. *Statistical Distributions in Scientific Work*, Vol. 4. C. Taillie et al., Eds. Reidel, Dordrecht, pp. 321–329.

Kemp, C. D. and Loukas, S. (1978) The computer generation of bivariate discrete random variables. *J. Roy. Statist. Soc.* A **141**, 513–519.

Kemp, C. D. and Loukas, S. (1981) Fast methods for generating bivariate discrete random variables. *Statistical Distributions in Scientific Work*, Vol. 4. C. Taillie et al., Eds. Reidel, Dordrecht, pp. 313–319.

Kendall, M. G. and Moran, P. A. P. (1963) *Geometrical Probability.* Griffin, London.

Kiefer, J. and Wolfowitz, J. (1952) Stochastic estimation of the maximum of a regression function. *Ann. Math. Statist.* **23**, 462–466.

Kinderman, A. J. and Monahan, J. F. (1977) Computer generation of random variables using the ratio of uniform deviates. *ACM Trans. Math. Soft.* **3**, 257–260.

Kinderman, A. J. and Monahan, J. F. (1980) New methods for generating Student's *t* and gamma variables. *Computing* **25**, 369–377.

Kinderman, A. J. and Ramage, J. G. (1976) Computer generation of normal random variables. *J. Amer. Statist. Assoc.* **71**, 893–896.

Kinderman, A. J., Monahan, J. F., and Ramage, J. G. (1977) Computer methods for sampling from Student's *t* distribution. *Math. Comp.* **31**, 1009–1018.

Kirkpatrick, S. Gelatt, C. D., Jr., and Vecchi, M. P. (1983) Optimization by simulated annealing. *Science* **220**, 671–680.

Kleijnen, J. P. C. (1974/5) *Statistical Techniques in Simulation.* Parts 1 and 2. Marcel Dekker, New York.

Kleijnen, J. P. C., van der Ven, R., and Sanders, B. (1982) Testing independence of simulation subruns: a note on the power of the von Neumann test. *Euro. J. Oper. Res.* **9**, 92–93.

Knuth, D. E. (1973a) *The Art of Computer Programming. Volume 1: Fundamental Algorithms.* 2nd ed. Addison-Wesley, Reading, MA.

Knuth, D. E. (1973b) *The Art of Computer Programming. Volume 3: Sorting and Searching.* Addison-Wesley, Reading, MA.

Knuth, D. E. (1981) *The Art of Computer Programming. Volume 2: Seminumerical Algorithms.* 2nd ed. Addison-Wesley, Reading, MA.

Knuth, D. E. (1984) An algorithm for Brownian zeroes. *Computing* **33**, 89–94.

Koksma, J. H. (1942) A general theorem from the theory of uniform distribution modulo 1 (in Dutch). *Mathematika Zutphen* **B11**, 7–11.

Kolmogorov, A. N. (1963) On tables of random numbers. *Sankhya* A **25**, 369–376.

Korobov, N. M. (1959) The approximate calculation of multiple integrals. *Dokl. Akad. Nauk. SSR* **124**, 1207–1210.

Kronmal, R. A. and Peterson, A. V., Jr. (1979) On the alias method for generating random variables from a discrete distribution. *Amer. Statist.* **33**, 214–218.

Kronmal, R. A. and Peterson, A. V. Jr. (1981) A variant of the acceptance-rejection method for the computer generation of random variables. *J. Amer. Statist. Assoc.* **76**, 446–451.

Kronmal, R. A. and Peterson, A. V. Jr. (1984) An acceptance-complement analogue of the mixture-plus-acceptance-rejection method for generating random variables. *ACM Trans. Math. Soft.* **10**, 271–281.

Lavenberg, S. S. and Sauer, C. H. (1977) Sequential stopping rules for the regenerative method of simulation. *IBM J. Res. Dev.* **21**, 545–558.

Lavenberg, S. S. and Welch, P. D. (1981) A perspective on the use of control variables to increase the efficiency of Monte Carlo simulations. *Manag. Sci.* **27**, 322–335.

Lavenberg, S. S., Moeller, T. L., and Welch, P. D. (1982) Statistical results on control variables with application to queueing network simulation. *Oper. Res.* **30**, 182–202.

Lehmer, D. H. (1951) Mathematical methods in large-scale computing units. *Proceedings of the Second Symposium on Large-Scale Digital Calculating Machinery.* Harvard University Press, Cambridge, MA, pp. 141–146.

Lewis, P. A. W. and Shedler, G. S. (1976) Simulation of non-homogeneous Poisson processes with log-linear rate function. *Biometrika* **63**, 501–505.

Lewis, P. A. W. and Shedler, G. S. (1979a) Simulation of non-homogeneous Poisson processes by thinning. *Nav. Res. Log. Q.* **26**, 403–413.

Lewis, P. A. W. and Shedler, G. S. (1979b) Simulation of non-homogeneous Poisson processes with degree-two exponential polynomial rate function. *Oper. Res.* **27**, 1026–1041.

Lewis, P. A. W., Goodman, A. S., and Miller, J. M. (1969) A pseudo-random number generator for the System/360. *IBM Sys. J.* **8**, 136–145.

Lewis, T. G. and Payne, W. H. (1973) Generalized feedback shift register pseudorandom number algorithms. *J. Assoc. Comput. Mach.* **20**, 456–468.

Lotwick, H. W. and Silverman, B. W. (1981) Convergence of spatial birth-and-death processes. *Math. Proc. Camb. Phil. Soc.* **90**, 155–165.

Loukas, S. and Kemp, C. D. (1983) On computer sampling from trivariate and multivariate discrete distributions. *J. Stat. Comp. Siml.* **17**, 113–123.

Lurie, D. and Hartley, H. O. (1972) Machine generation of order statistics for Monte Carlo computations. *Amer. Statist.* **26**, 26–27.

Lurie, D. and Mason, R. L. (1973) Empirical investigation of general techniques for computer generation of order statistics. *Comm. Statist.* **2**, 363–371.

MacLaren, M. D. and Marsaglia, G. (1965) Uniform random number generators. *J. Assoc. Comput. Mach.* **12**, 83–89.

McLeod, A. I. and Bellhouse, D. R. (1983) A convenient algorithm for drawing a simple random sample. *Appl. Statist.* **32**, 182–184.

Mantel, N. (1953) An extension of the Buffon needle problem. *Ann. Math. Statist.* **24**, 674–677.

Marriott, F. H. C. (1979) Barnard's Monte Carlo tests: how may simulations? *Appl. Statist.* **28**, 75–77.

Marsaglia, G. (1961a) Expressing a random variable in terms of uniform random variables. *Ann. Math. Statist.* **32**, 894–899.

Marsaglia, G. (1961b) Generating exponential random variables. *Ann. Math. Statist.* **32**, 899–900.

Marsaglia, G. (1963) Generating discrete random variables in a computer. *Comm. Assoc. Comput. Mach.* **6**, , 37–38.

Marsaglia, G. (1964) Generating a variable from the tail of a normal distribution. *Technometrics* **6**, 101–102.

Marsaglia, G. (1968) Random numbers fall mainly in the planes. *Proc. Nat. Acad. Sci. USA* **61**, 25–28.

Marsaglia, G. (1972) The structure of linear congruential sequences. *Applications of Number Theory to Numerical Analysis.* S. K. Zaremba, Ed. Academic Press, London, pp. 249–285.

Marsaglia, G. (1977) The squeeze method for generating gamma variates. *Comp. Math. Appl.* **3**, 321–325.

Marsaglia, G. (1980) Generating random variables with a *t*-distribution. *Math. Comp.* **34**, 235–236.

Marsaglia, G. and Bray, T. A. (1964) A convenient method for generating normal variables. *SIAM Review* **6**, 260–264.

Marsaglia, G., Ananthanarayanan, K., and Paul, N. J. (1976) Improvements on fast methods for generating normal random variables. *Inf. Proc. Lett.* **5**, 27–30.

Marsaglia, G., MacLaren, M. D., and Bray, T. A. (1964) A fast procedure for generating normal random variables. *Comm. Assoc. Comput. Mach.* **7**, 4–10.

Martin-Löf, P. (1966) The definition of random sequences. *Inf. Control* **9**, 602–619.

Martin-Löf, P. (1969) Algorithms and randomness. *Rev. Int. Statist. Inst.* **37**, 265–272.

Matheron, G. (1973) The intrinsic random functions and their applications. *Adv. Appl. Probab.* **5**, 439–468.

Meketon, M. S. and Heidelberger, P. (1982) A renewal theoretic approach to bias reduction in regenerative simulation. *Manag. Sci.* **28**, 173–181.

Metropolis, N., Rosenbluth, A. W., Rosenbluth, M. N., Teller, A. H., and Teller, E. (1953) Equations of state calculations by fast computing machines. *J. Chem. Phys.* **21**, 1087–1092.

Michael, J. R., Schucany, W. R., and Haas, R. W. (1976) Generating random variates using transformations with multiple roots. *Amer. Statist.* **30**, 88–90.

von Mises, R. (1919) Grundlagen der Wahrscheinlichkeitsrechnung. *Math. Zeit.* **5**, 52–99.

von Mises, R. (1957) *Probability, Statistics, Truth*. Macmillan, New York.

Mitchell, B. (1973) Variance reduction by antithetic variates in GI/G/1 queueing simlations. *Oper. Res.* **21**, 988–997.

Moran, P. A. P. (1975) The estimation of standard errors in Monte Carlo simulation experiments. *Biometrika* **62**, 1–4.

Moses, L. E. and Oakford, R. V. (1963) *Tables of Random Permutations*. Stanford University Press, Stanford, CA.

Nance, R. E. and Overstreet, C. L., Jr. (1972) A bibliography on random number generators. *Comp. Rev.* **13**, 495–508.

Nance, R. E. and Overstreet, C. L., Jr. (1978) Some experimental observations on the behavior of composite random number generators. *Oper. Res.* **26**, 915–935.

Nash, J. C. (1979) *Compact Numerical Methods for Computers*. Adam Hilger, Bristol.

Neave, H. R. (1973) On using the Box-Muller transformation with multiplicative congruential pseudo-random number generators. *Appl. Statist.* **22**, 92–97.

von Neumann, J. (1951) Various techniques in connection with random digits. *NBS Appl. Math. Ser.* **12**, 36–38.

Neveu, J. (1965) *Mathematical Foundations of the Calculus of Probabilities*. Holden-Day, San Fransisco.

Newby, M. J. (1979) The simulation of order statistics from life distributions. *Appl. Statist.* **28**, 298–301.

Newell, G. F. (1982) *Applications of Queueing Theory*, 2nd ed. Chapman and Hall, London.

Niederreiter, H. (1978) Quasi-Monte Carlo methods and pseudo-random numbers. *Bull. Amer. Math. Soc.* **84**, 957–1041.

Norman, J. E. and Cannon, L. E. (1972) A computer program for the generation of random variables from any discrete distribution. *J. Statist. Comput. Siml.* **1**, 331–348.

Odell, P. L. and Feiveson, A. H. (1966) A numerical procedure to generate a sample covariance matrix. *J. Amer. Statist. Assoc.* **61**, 199–203.

Page, E. S. (1965) On Monte Carlo methods in congestion problems II—simulation of queueing systems. *Oper. Res.* **13**, 300–305.

Page, E. S. (1967) A note on generating random permutations. *Appl. Statist.* **16**, 273–274.

Payne, W. H. (1970) Fortran Tausworthe pseudorandom number generator. *Comm. Assoc. Comput. Mach.* **13**, 57.

Payne, W. H., Rabung, J. H. and Bogyo, T. P. (1969) Coding the Lehmer pseudorandom number generator. *Comm. Assoc. Comput. Mach.* **12**, 85–86.

Perlman, M. D. and Wichura, M. J. (1975) Sharpening Buffon's needle. *Amer. Statist.* **29**, 157–163.

Peskun, P. H. (1973) Optimal Monte-Carlo sampling using Markov chains. *Biometrika* **60**, 607–612.

Peterson, A. V. Jr and Kronmal, R. A. (1983) Analytic comparison of three general-purpose methods for the computer generation of discrete random variables. *Appl. Statist.* **32**, 276–286.

Pincus, M. (1968) A closed form solution of certain programming problems. *Oper. Res.* **16**, 690–694.

Pincus, M. (1970) A Monte-Carlo method for the approximate solution of certain types of constrained optimization problems. *Oper. Res.* **18**, 1225–1228.

Platen, E. (1981) An approximation method for a class of Itô processes. *Lithuanian Math. J.* XXI, 121–133.

Preston, C. (1977) Spatial birth-and-death processes. *Bull. Int. Statist. Inst.* **46**(2), 371–391.

Priestley, M. B. (1981) *Spectral Analysis and Time Series.* Academic Press, London.

Rabinowitz, M. and Berenson, M. L. (1974) A comparison of various methods of obtaining random order statistics for Monte-Carlo computation. *Amer. Statist.* **28**, 27–29.

Ramberg, J. S. and Tadikamalla, P. R. (1978) On generation of subsets of order statistics. *J. Stat. Comp. Siml.* **6**, 239–241.

RAND Corporation (1955) *A Million Random Digits with 100,000 Normal Deviates.* Free Press, Glencoe, IL.

Rao, N. J., Borowankar, J. D., and Ramakrishna, D. (1974) Numerical solution of Itô integral equations. *SIAM J. Control* **12**, 124–139.

Reeder, H. A. (1972) Machine generation of order statistics. *Amer. Statist.* **26**, 56–57.

Relles, D. A. (1970) Variance reduction techniques for Monte Carlo sampling from Student distributions. *Technometrics* **12**, 499–515.

Relles, D. A. (1972) A simple algorithm for generating binomial random variables when N is large. *J. Amer. Statist. Assoc.* **67**, 612–613.

Research Machines Ltd. (1982) Letter to the author.

Richards, M. and Whitby-Strevens, C. (1979) *BCPL–The Language and its Compiler.* Cambridge University Press, London.

Ripley, B. D. (1977) Modelling spatial patterns (with discussion). *J. Roy. Statist. Soc.* B **39**, 172–212.

Ripley, B. D. (1979) Algorithm AS137. Simulating spatial patterns: dependent samples from a multivariate density. *Appl. Statist.* **28**, 109–112.

Ripley, B. D. (1981) *Spatial Statistics.* Wiley, New York.

Ripley, B. D. (1983a) The lattice structure of pseudo-random number generators. *Proc. Roy. Soc.* A **389**, 197–204.

Ripley, B. D. (1983b) Take your pick. *Personal Computer World* Sept. 1983, 188–191.

Ripley, B. D. (1983c) Computer generation of random variables: a tutorial. *Int. Statist. Rev.* **51**, 301–319.

Ripley, B. D. and Silverman, B. W. (1978) Quick tests for spatial interaction. *Biometrika* **65**, 641–642.

Robbins, H. and Monro, S. (1951) A stochastic approximation method. *Ann. Math. Statist.* **22**, 400–407.

Rosenblatt, M. (1975) Multiply schemes and shuffling. *Math. Comp.* **29**, 929–934.

Ross, S. M. (1970) *Applied Probability Models with Optimization Applications.* Holden-Day, San Francisco.

Rotenberg, A. (1960) A new pseudo-random number generator. *J. Assoc. Comput. Mach.* **7**, 75–77.

Rothery, P. (1982) The uses of control variates in Monte Carlo estimation of power. *Appl. Statist.* **31**, 125–129.

Rubinovitch, M. (1985) The slow server problem. *J. Appl. Probab.* **22**, 205–213.

Ruppert, D., Reisch, R. L., Deriso, R. B., and Carroll, R. J. (1984) Optimization using stochastic approximation and Monte Carlo simulation (with application to the harvesting of Atlantic menhaden). *Biometrics* **40**, 535–545.

Sahai, H. (1979) A supplement to Sowey's bibliography on random number generation and related topics. *J. Stat. Comp. Siml.* **10**, 31–52.

Sakesegawa, H. (1983) Stratified rejection and squeeze method for generating beta random numbers. *Ann. Inst. Statist. Math.* **35B**, 291–302.

Scheur, E. M. and Stoller, D. S. (1962) On the generation of normal random vectors. *Technometrics* **4**, 278–281.

Schmeiser, B. W. (1982) Batch size effects in the analysis of simulation output. *Oper. Res.* **30**, 556–568.

Schmeiser, B. W. and Babu, A. J. G. (1980) Beta variate generation via exponential majorizing functions. *Oper. Res.* **28**, 917–926.

Schmeiser, B. W. and Lal, R. (1980) Squeeze methods for generating gamma variates. *J. Amer. Statist. Assoc.* **75**, 679–682.

Schmeiser, B. W. and Shalaby, M. A. (1980) Acceptance/rejection methods for beta variate generation. *J. Amer. Statist. Assoc.* **75**, 673–678.

Schruben, L. W. (1982) Detecting initialization bias in simulation output. *Oper. Res.* **30**, 569–590.

Schruben, L. W. (1983) Confidence interval estimation using standardized time series. *Oper. Res.* **31**, 1090–1108.

Schruben, L. W. and Margolin, B. H. (1978) Pseudo-random number assignment in statistically designed simulation and distribution sampling experiments. *J. Amer. Statist. Assoc.* **73**, 504–525.

Schruben, L. W., Singh, H., and Tierney, L. (1983) Optimal tests for initialization bias in simulation output. *Oper. Res.* **31**, 1167–1178.

Schucany, W. R. (1972) Order statistics in simulation. *J. Stat. Comp. Siml.* **1**, 281–286.

Schuster, E. F. (1974) Buffon's needle experiment. *Amer. Math. Monthly* **81**, 26–29.

Seila, A. F. (1982) A batching approach to quantile estimation in regenerative simulations. *Manag. Sci.* **28**, 573–581.

Shedler, G. S. and Southard, J. (1982) Regenerative simulation of networks of queues with general service times: passage through subnetworks. *IBM J. Res. Dev.* **26**, 625–633.

Sibson, R. (1984) Personal communication.

Siegmund, D. (1976) Importance sampling in the Monte Carlo study of sequential tests. *Ann. Statist.* **4**, 673–684.

Simon, G. (1976) Computer simulation swindles, with applications to estimates of location and dispersion. *Appl. Statist.* **25**, 266–274.

Sironvalle, M. A. (1980) The random coin method: solution of the problem of simulation of a random function in the plane. *Math. Geol.* **12**, 25–32.

Smith, C. S. (1971) Multiplicative pseudo-random number generators with prime modulus. *J. Assoc. Comput. Mach.* **18**, 587–593.

Smith, W. B. and Hocking, R. R. (1972) Algorithm AS53. Wishart variate generator. *Appl. Statist.* **21**, 341–345.

Solomon, H. (1978) *Geometric Probability.* SIAM, Philadelphia.

Sowey, E. R. (1972) A chronological and classified bibliography on random number generation and testing. *Int. Statist. Rev.* **40**, 355–371.

Sowey, E. R. (1978) A second classified bibliography on random number generation and testing. *Int. Statist. Rev.* **46**, 89–102.

Springer, B. G. F. (1969) Numerical optimization in the presence of random variability. The single factor case. *Biometrika* **56**, 65–74.

"Student" (1908) The probable error of a mean. *Biometrika* **6**, 1–25.

Swick, D. A. (1974) Letter to the editors. *Appl. Statist.* **23**, 233.

Tadikamalla, P. R. (1978) Computer generation of gamma random variables. *Comm. Assoc. Comput. Mach.* **21**, 419–423, 925–928.

Tadikamalla, P. R. (1979) Random sampling from the generalized gamma distribution. *Computing* **23**, 199–203.

Tadikamalla, P. R. and Johnson, M. E. (1981) A complete guide to gamma variate generation. *Amer. J. Math. Mang. Sci.* **1**, 213–236.

Tausworthe, R. C. (1965) Random numbers generated by linear recurrence modulo two. *Math. Comp.* **19**, 201–209.

Thompson, W. E. (1958) A modified congruence method of generating pseudo-random numbers. *Computer J.* **1**, 83, 86.

Thompson, W. E. (1959) ERNIE—a mathematical and statistical analysis. *J. Roy. Statist. Soc.* A **122**, 301–333.

Tippett, L. H. C. (1927) *Random Sampling Numbers.* Tracts for Computers XV. Cambridge University Press, London.

Tocher, K. D. (1954) The application of automatic computers to sampling experiments. *J. Roy. Statist. Soc.* B **16**, 39–61.

Tocher, K. D. (1963) *The Art of Simulation.* English Universities Press, London.

Tootill, A. (1982) PCW subset. *Personal Computer World* June, 133.

Tootill, J. P. F., Robinson, W. D., and Adams, A. G. (1971) The runs up-and-down performance of Tausworthe pseudo-random number generators. *J. Assoc. Comput. Mach.* **18**, 381–399.

Tootill, J. P. R., Robinson, W. D., and Eagle, D. J. (1973) An asymptotically random Tausworthe sequence. *J. Assoc. Comput. Mach.* **20**, 469–481.

Trotter, H. F. and Tukey, J. W. (1956) Conditional Monte Carlo for normal samples. *Symposium on Monte Carlo Methods.* H. A. Meyer, Ed. Wiley, New York, pp. 64–79.

Vitter, J. S. (1984) Faster methods for random sampling. *Comm. Assoc. Comput. Mach.* **27**, 703–718.

Walker, A. J. (1974) New fast method for generating discrete random variables with arbitrary frequency distribution. *Elec. Lett.* **10**, 127–128.

Walker, A. J. (1977) An efficient method for generating discrete random variables with general distributions. *ACM Trans. Math. Soft.* **3**, 253–256.

Wasan, M. T. (1969) *Stochastic Approximation.* Cambridge University Press, London.

Wasow, W. (1952) A note on the inversion of matrices by random walks *Math. Comp.* **6**, 78–81.

Welch, P. D. (1983) The statistical analysis of simulation results. *The Computer Performance Modeling Handbook.* S. S. Lavenberg, Ed. Academic Press, New York, pp. 268–328.

West, J. H. (1955) An analysis of 162,332 lottery numbers. *J. Roy. Statist. Soc.* A **118**, 417–426.

Whitt, W. (1976) Bivariate distributions with given marginals. *Ann. Statist.* **4**, 1280–1289.

Whittlesey, J. R. B. (1968) A comparison of the correlation behavior of random number generators for the IBM 360. *Comm. Assoc. Comp. Mach.* **11**, 641–644.

Whittlesey, J. R. B. (1969) On the multidimensional uniformity of pseudorandom generators. *Commun. Assoc. Comput. Mach.* **12**, 247.

Wichmann, B. A. and Hill, J. D. (1982) Algorithm AS183. An efficient and portable pseudo-random number generator. *Appl. Statist.* **31**, 188–190; **33**, 123.

Wilson, J. R. (1979) Proof of the antithetic-variates theorem for unbounded functions. *Math. Proc. Camb. Phil. Soc.* **86**, 477–479.

Wilson, J. R. and Pritsker, A. A. B. (1978) A survey of research on the simulation start-up problem. *Simulation* **31**, 55–58.

Wong, C. K. and Easton, M. C. (1980) An efficient method for weighted sampling without replacement. *SIAM J. Computing* **9**, 111–113.

Wright, R. D. and Ramsay, T. E., Jr. (1979) On the effectiveness of common random numbers. *Manag. Sci.* **25**, 649–656.

Zierler, N. (1969) Primitive polynomials whose degree is a Mersenne exponent. *Inf. Control* **15**, 67–69.

Zierler, N. and Brillhart, J. (1968) On primitive trinomials (mod 2). *Inf. Control* **13**, 541–554.

Zierler, N. and Brillhart, J. (1969) On primitive trinomials (mod 2) II. *Inf. Control* **14**, 566–569.

Zubrzycki, S. (1957) On estimating gangue parameters (in Polish). *Zastos. Mat.* **3**, 105–153.

APPENDIX A

Computer Systems

The examples were computed on a variety of machines ranging from personal computers to scientific mainframes. The details given below may help in understanding the timings quoted.

BBC Microcomputer

A personal computer manufactured by Acorn Computers under licence to the British Broadcasting Corporation and widely used in education in the UK. It is based on the 6502 8-bit microprocessor running at 2 MHz. The BASIC interpreter supplied was used. This is an advanced Basic with repeat...until loops and recursive procedures and functions. The intrinsic RND pseudo-random function is based on a Tausworthe generator. (See Section 2.3 for a description.) The real variables are contained in 5 bytes with a 32-bit precision. Integer variables are 32 bits long with range $-2^{31} \cdots 2^{31} - 1$.

ACT Sirius 1

A business microcomputer, very similar to the Victor 9000 marketed in North America. It is based on the 8088 8/16-bit microprocessor. The Microsoft BASIC interpreter and compiler were run under MS-DOS. The inbuilt (and unspecified) pseudo-random function was used. The real variables are contained in 4 bytes with 23-bit precision.

Corvus Concept

A workstation based on the 68000 16/32-bit microprocessor running at 8 MHz with wait states. The SVS Fortran 77 compiler was used. Real variables are 4 bytes long with 24-bit precision; double precision variables are 8 bytes long with 53-bit precision. Both conform to the IEEE standard (*Computing*, March 1981). Integer variables are 32 bits long with range $-2^{31} \cdots 2^{31} - 1$.

The congruential generator

$$X_i = (69096X_{i-1} + 1) \bmod 2^{32}$$

was coded in assembler using only the registers.

DEC VAX 11/782

A dual-processor 32-bit superminicomputer running the VAX/VMS operating system. Timings quoted are CPU time for programs compiled under the optimizing Fortran 77 compiler, and would be similar for a VAX 11/780. The congruential generator $X_i = (69069X_{i-1} + 1) \bmod 2^{32}$ is supplied and was used. Real variables have 24-bit precision and double precision variables have 56-bit precision. Integer variables are again 32 bits with range $-2^{31} \ldots 2^{31} - 1$.

CDC Cyber 174

A 60-bit scientific mainframe with an optimizing Fortran 77 compiler used at full optimization. Timings quoted are CPU time. The pseudo-random function used was either the intrinsic function RANF(), which implements

$$X_i = 44,485,709,377,909 X_{i-1} \bmod 2^{48}$$

or the function G05CAF from the NAG library which implements

$$X_i = 5^{13} X_{i-1} \bmod 2^{59}$$

Real variables are 60 bits long with 48-bit precision, whereas double precision variables have 96-bit accuracy. Integers are stored in 60 bits, but most operations are restricted to magnitudes less than 2^{48}.

APPENDIX B

Computer Programs

The programs included here are intended to show to avoid some of the pitfalls in implementing the algorithms of Chapters 2 and 3. They are written in Fortran 77 as the only widely available language with an extended precision data type. Let *maxint* be the largest integer such that all integers with modulus up to and including *maxint* are represented exactly in double precision variables.

B.1. FORM $a \times b \bmod c$

This is surprisingly difficult to do with adequate generality. The program given here is fairly slow, but is useful to check the operation of special-purpose code, for example, when implementing congruential generators.

The problems stem from the limited range of the integer type, which typically only represents integers up to $2^{31} - 1$, and sometimes $2^{15} - 1$. Double precision variables will represent larger integers exactly, with $maxint = 2^{53}$ and 2^{56} on the Corvus and Vax machines described in Appendix A. There will be no warning of loss of accuracy when using double precision variables, so care is needed to ensure that integer terms never exceed *maxint* in modulus. If a higher precision type than double precision is available it can be substituted. There is still a problem, for the INT function on double precision variables is often restricted to integer parts $\leqslant 2^{31} - 1$, so this function has to be avoided together with the MOD function.

The function MUL (A, B, C) computes $A \times B \bmod C$ for integers A, B, C, with $0 \leqslant A$, $B < C \leqslant M^2$, where $2M^2 \leqslant maxint$ and $INT(2*M - 1)$ is acceptable.

```
      FUNCTION MUL(A, B, C)
C     forms A*B MOD C for A, B < C < = 2^50
      DOUBLE PRECISION A, B, C, MUL, A1, A2, B1, B2, D, M, MM
```

```
C       set M so that 2•M•M < = maxint
        M = 2.0D0••25
        MM = M•M
        D = (A•B)/C
        CALL SPLIT (A, B, A1, A2)
        CALL SPLIT (C, D, B1, B2)
        MUL = A1 - B1 + (A2 - B2)•MM
        RETURN
        END
C
        SUBROUTINE SPLIT (A, B, C, D)
C       forms A•B = C + D•MM for 0 < = C, D < MM
        DOUBLE PRECISION A, B, C, D, A1, A2, B1, B2, M, MM, AC, AD, C1, C2
C       set M to the same value as in MUL
        M = 2.0D0••25
        MM = M•M
        A2 = INT(A/M)
        A1 = INT(A - A2•M)
        B2 = INT(B/M)
        B1 = INT(B - B2•M)
        AC = A2•B1 + A1•B2
        C2 = INT(AC/M)
        C1 = AC - M•C2
        AC = C1•M + A1•B1
        AD = INT(AC/MM)
        C = AC - AD•MM
        D = AD + C2 + A2•B2
        RETURN
        END
```

The subroutine SPLIT sets $int(A) = A1 + A2 * M$, $int(B) = B1 + B2 * M$, so

$$int(A) * int(B) = A1 * B1 + AC * M + A2 * B2 * M^2$$

However, $0 \leq AC < 2M^2$, so we set $AC = C1 + C2 * M$, where $0 \leq C1 < M$. Then

$$int(A) * int(B) = (A1 * B1 + C1 * M) + (A2 * B2 + C2) * M^2$$

Again, the first term might exceed M^2, so we write this as $C + AD * M^2$, to obtain the decomposition

$$int(A) * int(B) = C + D * M^2$$

We know $A, B < M^2$, so $D < M^2$.

The function MUL sets $D = (A * B)/C$, so at least int(D) will be accurate. Then SPLIT gives

$$A * B = A1 + A2 * N^2$$

$$C * int(D) = B1 + B2 * M^2$$

so the answer is the difference $(A1 - B1) + (A2 - B2) * M^2$. We know this to be less than C, so $(A2 - B2)$ is zero or one.

Program B.3 contains a modification of this program with SPLIT forming nint(A) * nint(B). This forms $A * B$ mod C with the residue in the range $-C/2, \ldots, C/2$. (If C is even, which extreme occurs depends on the sign of $A * B$.)

B.2. CHECK PRIMITIVE ROOTS

To ascertain whether a is a primitive root modulo M we need to know the prime factorization

$$M - 1 = p_1^{x_1} \cdots p_r^{x_r}$$

[Knuth (1981, Section 4.5.4) discusses how to find such factorizations.] Then we check that $a^{(M-1)/p_i} \not\equiv 1$ mod M for $i = 1, \ldots, r$. The program PROOT does this for $2M \leqslant maxint, 2^{51}$. Useful test cases are $M = 2^{31} - 1 = 2147483647$ with $M - 1 = 2.3^2.7.11.31.151.331$ for which 7 and $7^5 = 16807$ are primitive roots, and 13 is not.

```
        PROGRAM PROOT
        DOUBLE PRECISION A(50), AA, FAC, M, MUL, MULT, PI, PWR
        INTEGER I, J, R
        PRINT *, 'Multiplier, Modulus, No of factors of M − 1'
        READ *, MULT, M, R
        AA = MULT
        A(1) = MULT
        DO 10 I = 2, 50
          AA = MUL (AA, AA, M)
 10       A(I) = AA
        DO 30 I = 1, R
          PRINT *, 'Factor', I
          READ *, FAC
          PWR = (M − 1)/FAC
          AA = 1.0D0
          DO 20 J = 50, 1, −1
```

```
          PI = 2.0D0**(J-1)
          IF (PWR. GE. PI) THEN
             PWR = PWR - PI
             AA = MUL (AA, A(J), M)
          ENDIF
    20    CONTINUE
          IF (AA. EQ. 1.0D0) THEN
             PRINT *, 'FAILED'
          ELSE
             PRINT *, 'OK for this factor'
          ENDIF
    30    CONTINUE
          END
```

The function MUL is given in B.1. The key identity used is that

$$ab \bmod M = (a \bmod M)(b \bmod M)\bmod M$$

[Let $a = a' + a''M$, $b = b' + b''M$. Then $ab = a'b' + (a'b'' + a''b' + a''b''M)M$ so $ab \bmod M = a'b' \bmod M$.] This is used to precompute

$$A(i + 1) = a^{2^i} \bmod M, \qquad i = 0, \dots, 49$$

For any integer $t, 0 \leqslant t < 2^{50}$, its binary representation $t_{49} \cdots t_0$ is found, and

$$a^t \bmod M = \left[\prod_{t_i = 1} (a^{2^i} \bmod M) \right] \bmod M$$

is computed. The restrictions arise from MUL and the need for all $(M - 1)/p_i < 2^{50}$.

B.3. LATTICE CONSTANTS FOR CONGRUENTIAL GENERATORS

Suppose $X_i = (aX_{i-1} + c) \bmod M$ is a congruential generator of full or maximal period. The subroutine LATT computes the lattice constants r, l_k, and v_k discussed in Section 2.4 for dimensions $k = 2, \dots, R$.

Double precision variables are used throughout for greatest accuracy. The rows of X are the basis vectors Me_i. Throughout powers of a are reduced modulo M to $-M/2, \dots, M/2$. Thus initially the length of Me_1 is at most $1 + (k - 1) \times (M/2)^2 < 2M$ for $k \leqslant 8$. At all times the length of the vectors is reduced, so we can guarantee that no element of X will ever exceed $2M$ in magnitude for $k \leqslant 8$. Internal terms in TEST1 and TEST2 are bounded by four times the bound on an element of X. Thus we are guaranteed accuracy if integers up to $8M$ are representable exactly. This is very much a worst case, extremely unlikely to occur as the elements of X normally reduce in size to $O(\sqrt{M})$ after a few steps of TEST1. Warnings are issued if problems might occur.

The polar basis $\{e_i^*\}$ is found by inverting the matrix of $\{e_i\}$ using Gaussian elimination with partial pivoting. It is possible to update the two bases simultaneously as illustrated in Section 2.4. However, the elements of $\{e_i^*\}$ can become larger than M and accuracy may be lost.

Entry parameters

> R integer max dimension $\leqslant 8$
>
> MULT double precision multiplier a
>
> M double precision modulus M

One also needs to set M in MUL and SPLIT so that $2M^2 \leqslant maxint$, and MX in TEST1 and TEST2 to $maxint$.

```
            SUBROUTINE LATT (R, MULT, M)
            LOGICAL TEST2, TEST1
            INTEGER R, K
            DOUBLE PRECISION X(8, 8), LEN(8), MULT, M
            DO 20 K = 2, R
               CALL INIT (X, LEN, K, MULT, M)
      10       IF (TEST1 (X, LEN, K)) GO TO 10
               IF (TEST2 (X, LEN, K)) GO TO 10
               CALL RES (X, LEN, K, M)
      20    CONTINUE
            RETURN
            END
   C
            SUBROUTINE INIT (X, LEN, R, MULT, M)
            INTEGER R, I, K
            DOUBLE PRECISION X(8, 8), LEN(8), L, MULT, M, A, MUL
            IF (R. EQ. 2) THEN
            L = 1.0
            DO 10 K = 1, 8
               X(1, K) = L
      10       L = MUL(L, MULT, M)
            DO 20 I = 2, 8
               DO 20 K = 1, 8
                  X(I, K) = 0.0
                  IF (I.EQ. K) X(I, K) = M
```

```
20          CONTINUE
       ELSE
          A = X(1, R)
          DO 25 I = 1, R - 1
25          X(I, R) = MUL(A, X(I, 1), M)
       ENDIF
       DO 40 I = 1, R
          A = 0.0
          DO 30 K = 1, R
30          A = A + X(I, K)•X(I, K)
40       LEN(I) = A
       RETURN
       END


       FUNCTION TEST1 (X, LEN, R)
       LOGICAL TEST1
       INTEGER R, I, J, K, I1, I2, L, NCHNGS
       DOUBLE PRECISION X(8, 8), LEN(8), XY, A, B, MX
       DATA MX/maxint/
       NCHNGS = 0
C      try each pair I, J in turn
       DO 40 I = 2, R
          DO 40 J = 1, I - 1
10          XY = 0.0
             DO 20 K = 1, R
20             XY = XY + X(I, K)•X(J, K)
             IF (LEN(I). LE. LEN(J)) THEN
                I1 = I
                I2 = J
             ELSE
                I1 = J
                I2 = I
             ENDIF
C      round halves towards zero
             A = XY/LEN(I1)
             L = INT(ABS(A) + 0.499999999)
             IF (A. LT. 0.0) L = -L
             IF (L. EQ. 0) GO TO 40
             NCHNGS = NCHNGS + 1
             A = 0.0
             DO 30 K = 1, R
               B = L•X(I1, K)
               IF (ABS(B). GT. MX) PRINT •, 'Accuracy loss'
               B = X(I2, K) - B
               X(I2, K) = B
30             A = A + B•B
             LEN(I2) ⩴ A
             IF (LEN(I2). LT. LEN(I1)) GO TO 10
40       CONTINUE
       TEST1 = NCHNGS. GT. 0
       RETURN
       END
```

```
            FUNCTION TEST2 (X, LEN, R)
            LOGICAL TEST2
            INTEGER I, I1, J, K, L, L1, R, S, S1, T1, CON(3), IN(8), PTR(8)
            DOUBLE PRECISION X(8, 8), LEN(8), T(8), A, AL, B, MX
            DATA CON/0, -1, +1/
            DATA MX/maxint/
            TEST2 = .FALSE.
            IF (R .EQ. 2) RETURN
            DO 10 I = 1, R
      10      PTR(I) = I
C           PTR is pointer to vectors in length order
            DO 30 I = 1, R - 1
              L = I
              AL = LEN(PTR(L))
              DO 20 J = I + 1, R
              L1 = PTR(J)
                IF (LEN(L1) .LT. AL) THEN
                    L = J
                    AL = LEN(L1)
                ENDIF
      20        CONTINUE
              IF (L .NE. I) THEN
                  L1 = PTR(L)
                  PTR(L) = PTR(I)
                  PTR(I) = L1
              ENDIF
      30      CONTINUE
C           try Minkowski's test
            DO 100 I = 3, R
              I1 = PTR(I)
              AL = LEN(I1) - 0.5
              IN(I) = 1
              DO 90 S = 1, 3**(I - 1) - 1
                S1 = S
                DO 50 L = 1, I - 1
                  T1 = S1/3
                  IN(L) = CON(S1 - 3*T1 + 1)
      50          S1 = T1
                A = 0. 0
                DO 70 J = 1, R
                  B = 0.0
                  DO 60 K = 1, I
                    B = B + X(PTR(K), J)*IN(K)
                    IF (ABS(B) .GT. MX) PRINT *, 'Possible accuracy loss'
      60            CONTINUE
                  T(J) = B
      70          A = A + B*B
                IF (A .GT. AL) GO TO 90
                LEN(PTR(I)) = A
                DO 80 J = 1, R
```

```
80          X(I1, J) = T(J)
            TEST2 = .TRUE.
            PRINT *, 'TEST2 SUCCESS'
            RETURN
90          CONTINUE
100      CONTINUE
         RETURN
         END


         SUBROUTINE RES (X, LEN, R, M)
         DOUBLE PRECISION X(8, 8), LEN(8), A(8, 8), M, LL, LU
         REAL NU, RATIO, AA, B, UL
         INTEGER R, Z(8), I, J
         LL = LEN(1)
         LU = LL
         DO 10 I = 2, R
            LL = MIN (LL, LEN(I))
10          LU = MAX (LU, LEN(I))
         RATIO = SQRT(LU/LL)
         NU = SQRT(LL)
         IF (R .EQ. 2) GO TO 50
         CALL INV (X, A, R, M)
         B = 1.0E20
         DO 30 I = 1, R
            AA = 0.0
            DO 20 J = 1, R
20             AA = AA + A(J, I)*A(J, I)
30          B = MIN(AA, B)
         NU = SQRT(B)
         DO 40 I = 1,R
40          Z(I) = INT(NU*SQRT(LEN(I))/M)
         CALL SEARCH(A, R, NU, Z, M)
50       UL = SQRT(LU)/M
         PRINT 1000, R, RATIO, UL, NU
1000     FORMAT (' DIM', I2, ' RATIO ', 1PG10.3, ' LMAX ', 1PE10.2,
         &' NU ', 1PE10.2)
         RETURN
         END


         SUBROUTINE INV (X, Y, R, M)
C        invert X/M to Y by Gaussian elimination
C        with partial pivoting
         INTEGER R, H, I, J, K, N
         DOUBLE PRECISION X(8, 8), Y(8, 8), M, W(8, 16), S
         N = R + R
         DO 20 I = 1, R
            DO 10 J = 1, R
10             W(I, J) = X(I, J)/M
            DO 20 J = 1, R
               IF (I .EQ. J) THEN
                  W(I, J + R) = 1.0
```

```
                      ELSE
                         W(I, J + R) = 0.0
                      ENDIF
     20               CONTINUE
              DO 60 J = 1, R − 1
                 S = ABS(W(J, J))
                 K = J
                 DO 30 H = J + 1, R
                    IF (ABS(W(H, J)) .GT. S) THEN
                       S = ABS(W(H, J))
                       K = H
                    ENDIF
     30          CONTINUE
                 IF (K .NE. J) THEN
                    DO 40 I = J, N
                       S = W(K, I)
                       W(K, I) = W(J, I)
     40                W(J, I) = S
                 ENDIF
                 DO 50 K = J + 1, R
                    W(K, J) = W(K, J)/W(J, J)
                    DO 50 I = J + 1, N
     50                W(K, I) = W(K, I) − W(K, J)•W(J, I)
     60       CONTINUE
              DO 90 I = R + 1, N
                 W(R, I) = W(R, I)/W(R, R)
                 DO 80 J = R − 1, 1, −1
                    S = W(J, I)
                    DO 70 K = J + 1, R
     70                S = S − W(J, K)•W(K, I)
     80             W(J, I) = S/W(J, J)
     90       CONTINUE
              DO 100 I = 1, R
                 DO 100 J = 1, R
     100           Y(I, J) = W(I, J + R)
              RETURN
              END


C

              SUBROUTINE SEARCH (A, R, NU, Z, M)
              INTEGER R, Z(8), T(8), I, J, K
              REAL NU, AC
              DOUBLE PRECISION M, A(8, 8), Y(8), AA
              DO 10 I = 1, R
                 T(I) = 0
     10
              K = R
     20    IF (T(K) .EQ. Z(K)) GO TO 80
              T(K) = T(K) + 1
              DO 30 J = 1, R
     30       Y(J) = Y(J) + A(J,K)
```

```
40   K=K+1
     IF (K .GT. R) GO TO 60
     T(K)=-Z(K)
     IF (Z(K) .NE.  0) THEN
        DO 50 J=1,R
50         Y(J)=Y(J)-2*Z(K)*A(J,K)
     ENDIF
     GO TO 40 ,
60   AA=0.0
     DO 70 J=1,R
70     AA=AA+Y(J)*Y(J)
     AC=SQRT(AA)
     IF (AC .LT. 0.9999*NU) THEN
        PRINT *, 'SEARCH SUCCESS'
        NU=AC
     ENDIF
80   K=K-1
     IF (K. GE. 1) GO TO 20
     RETURN
     END
C
C
     FUNCTION MUL(A, B, C)
C    forms A*B MOD C for A,B < C < =2^50
     DOUBLE PRECISION A,B,C,MUL,A1,A2,B1,B2,D,M,MM
C    set M so 2*M*M  < =maxint
     M=2. 0D0**25
     MM=M*M
     D=(A*B)/C
     CALL SPLIT (A,B,A1,A2)
     CALL SPLIT (C,D,B1,B2)
     MUL=A1-B1 +(A2-B2)*MM
     RETURN
     END
C
     SUBROUNTINE SPLIT (A,B,C,D)
     DOUBLE PRECISION A,B,C,D,A1,A2,B1,B2,M,MM,AC,AD,C1,C2
     M=2**25
     MM=M*M
     A2=INT(A/M)
     A1=NINT(A-A2*M)
     B2=INT(B/M)
     B1=NINT(B-B2*M)
     AC=A2*B1 +A1*B2
     C2=INT(AC/M)
     C1=AC-M*C2
     AC=C1*M +A1*B1
     AD=INT(AC/MM)
     C=AC-AD*MM
     D=AD +C2 +A2*B2
     RETURN
     END
```

B.4. TESTING GFSR GENERATORS

Theorem 2.9 provides a test of k-distribution for a GFSR generator with L-bit words (Y_i) and $kL \leqslant p$. The subroutine GFSRT takes the $p \times kL$ matrix with rows the bits of ($Y_i, \ldots, Y_{i+k-1}$), $i = 1, \ldots, p$ and reduces to upper triangular form following the example of Section 2.3.

```
        SUBROUTINE GFSRT (A,P,N)
C       set pmax as required
        INTEGER P,N,A(pmax,pmax),EOR,I,J,L,T
        DO 60 I=1,N
          J=I
          IF (A(I,I) .EQ. 0) THEN
10          J=J+1
            IF (A(J,I) .GT. 0) GO TO 20
            IF (J .LT. P) GO TO 10
            PRINT *, 'RANK DEFICIENT'
            RETURN
20          DO 30 L=I,N
              T=A(J,L)
              A(J,L)=A(I,L)
30            A(I,L)=T
          ENDIF
          DO 50 J=I+1,P
            IF (A(J,I) .GT. 0) THEN
              DO 40 L=I,N
40              A(J,L)=EOR(A(J,L),A(I,L))
            ENDIF
50        CONTINUE
60      CONTINUE
        PRINT *, 'FULL RANK'
        RETURN
        END

        FUNCTION EOR(I,J)
        INTEGER EOR, I,J
        EOR=MOD(I+J,2)
        RETURN
        END
```

Subroutine GFSRS prepares the matrix from $Y_1, \ldots, Y_{p+k-1}$.

```
        SUBROUTINE GFSRS (Y,L,P,K)
C       set pmax as required
        INTEGER L,P,K,Y(P+K-1),A(pmax,pmax),D,I1,J1,M,M1
        N=K*L
        IF (N .GT. P) THEN
          PRINT *, 'MUST FAIL'
          RETURN
```

```
          ENDIF
          M=2**(L-1)
          DO 30 I=1,P
            I1=0
            DO 20 J=1,K
              D=Y(I+J-1)
              M1=M
              DO 10 J1=1,L
                I1=I1+1
                IF (D .GE. M1) THEN
                  A(I,I1)=1
                  D=D-M1
                ELSE
                  A(I,I1)=0
                ENDIF
10                M1=M1/2
20            CONTINUE
30          CONTINUE
          CALL GFSRT (A,P,N)
          RETURN
          END
```

B.5. NORMAL VARIATES

In the remaining section RND( ) is the pseudo-random number generator.
The polar algorithm 3.5 requires one of its variates to be saved for a
subsequent call.

```
          FUNCTION POLAR ( )
          REAL AN,E,POLAR,VI,V2,W
          SAVE IR,AN
          DATA IR/0/
          IF (IR.EQ.0) THEN
10          V1=2.*RND( )-1.0
            V2=2.*RND( )-1.0
            W=V1*V1 +V2*V2
            IF (W.GT.1. ) GO TO 10
            E=SQRT((-2.0*LOG(W))/W)
            AN=V1*E
            IR=1
            POLAR=V2*E
          ELSE
            IR=0
            POLAR=AN
          ENDIF
          RETURN
          END
```

The SAVE statement is formally necessary to save IR and AN between calls. It is not usually necessary and may slow the program dramatically. Algorithm 3.17, the ratio method, does not have this problem.

```
      FUNCTION NRATIO( )
      REAL NRATIO,U,V,X,Z
  10  U=RND( )
      V=0.8578•(2.•RND( )-1.)
      X=V/U
      Z=0.25•X•X
      IF (Z .LT. 1.-U) GO TO 20
      IF (Z .GT. (0.259/U+0.35)) GO TO 10
      IF (Z .GT. -LOG(U)) GO TO 10
  20  NRATIO=X
      RETURN
      END
```

Algorithm 3.18, the Marsaglia–Bray composition method, is fast if pseudo-random numbers are cheap.

```
      FUNCTION NMB( )
      REAL NMB,AV,G,U,UI,U2,V,W
      U=RND( )
      IF (U.LE.0.8638) THEN
        NMB=2.0•(RND( )+RND( )+RND( )-1.5)
        RETURN
      ENDIF
      IF (U.LE.0.9745) THEN
        NMB=1.5•(RND( )+RND( )-1.0)
        RETURN
      ENDIF
      IF (U.LE.0.9973002039) THEN
  10    V=6.0•RND( )-3.0
        AV=ABS(V)
        G=17.49731196•EXP(-0.5•V•V)
        IF (AV .LT. 1.0) THEN
          G=G-4.73570326•(3.0-V•V)
        ELSE
          G=G-2.36785163•(3.0-AV)•(3.0-AV)
        ENDIF
        IF (AV .LT. 1.5) G=G-2.157875•(1.5-AV)
          IF (0.358•RND( ) .GT. G) GO TO 10
        NMB=V
        RETURN
      ENDIF
```

```
20   U1=2.0•RND( )−1.0
     U2=2.0•RND( )−1.0
     W=U1•U1+U2•U2
     IF (W .GE. 1.0) GO TO 20
     W=SQRT((9.0−2.0•LOG(W))/W)
     IF (ABS(U1•W) .GT. 3.0) THEN
        NMB=U1•W
        RETURN
     ENDIF
     IF (ABS(U2•W) .LE. 3.0) GO TO 20
     NMB=U2•W
     RETURN
     END
```

B.6.   EXPONENTIAL VARIATES

Algorithm 3.7 gives

```
     FUNCTION EXPRV( )
     REAL EXPRV,A,U,UO, USTAR
     A=0.0
10   U=RND( )
     UO=U
20   USTAR=RND( )
     IF (U .LT. USTAR) GO TO 30
     U=RND( )
     IF (U .LT. USTAR) GO TO 20
     A=A+1.0
     GO TO 10
30   EXPRV=A+UO
     RETURN
     END
```

B.7.   GAMMA VARIATES

For shape parameter $\alpha < 1$ we have Algorithm 3.19:

```
     FUNCTION GS (ALPHA)
     REAL GS,ALPHA,B,P,X
     DATA E/2.71828182/
     B=(ALPHA+E)/E
10   P=B•RND( )
     IF (P .GT.1.0) GO TO 20
     X=P••(1./ALPHA)
     IF (X .GT. −LOG(RND ( ))) GO TO 10
```

```
      GS=X
      RETURN
 20   X=-LOG((B-P)/ALPHA)
      IF (X**(ALPHA-1.0) .LT. RND( )) GO TO 10
      GS=X
      RETURN
      END
```

The case $\alpha = 1$ is the exponential dealt with in B.6. For $\alpha > 1$ we use Algorithm 3.20.

```
      FUNCTION GCF (ALPHA)
      REAL GCF,ALPHA,AA,C1,C2,C3,C4,C5,U1,U2,W,X
      SAVE APREV,C1,C2,C3,C4,C5
      DATA APREV/0.0/
      IF (ALPHA .EQ. APREV) GO TO 10
      C1=ALPHA-1.0
      AA=1.0/C1
      C2=AA*(ALPHA-1.0/(6.0*ALPHA))
      C3=2.0*AA
      C4=C3+2.0
      IF (ALPHA .GT. 2.5) C5=1.0/SQRT(ALPHA)
 10   U1=RND( )
      U2=RND( )
      IF (ALPHA .LE. 2.5) GO TO 20
      U1=U2+C5*(1.0-1.86*U1)
      IF (U1.LE.0.0 .OR. U1.GE. 1.0) GO TO 10
 20   W=C2*U2/U1
      IF (C3*U1+W+1.0/W .LT. C4) GO TO 30
      IF (C3*LOG(U1)-LOG(W)+W .GE. 1.0) GO TO 10
 30   GCF=C1*W
      APREV=ALPHA
      RETURN
      END
```

B.8. DISCRETE DISTRIBUTIONS

The two main contenders are an indexed search and the alias method.

For the indexed search let P be a table of cumulative probabilities, so $P(M) = 1.0$. Subroutine SETIND sets up an index array IND with MI entries, and DISRV uses this to generate samples.

```
      SUBROUTINE SETIND (P,M,IND,MI)
      REAL P(M),P0
      INTEGER IND(MI),M,MI,I,K
      I=1
      DO 20 K=1,MI
```

```
            PO=REAL(K−1)/REAL(MI)
   10       IF (P(I) .GE. PO) GO TO 20
            I=I+1
            GO TO 10
   20       IND(K)=I
            RETURN
            END


            FUNCTION DISRV(P,M,IND,MI)
            REAL P(M),U
            INTEGER DISRV, IND(MI),M,MI,I
            U=RND( )
            I=IND(INT(MI•U)+1)
   10       IF (P(I) .GE. U) GO TO 20
            I=I+1
            GO TO 10
   20       DISRV=I
            RETURN
            END
```

For the alias method we assume P contains the actual probabilities. Program SETAL forms the alias tables, using an integer workspace W of size at least M, and ALRV produces a sample. Algorithm 3.13B is used in SETAL.

```
            SUBROUTINE SETAL (P,M,A,Q,W)
            REAL P(M),Q(M)
            INTEGER M,A(M),W(M),I,J,NN,NP,S
            NN=0
            NP=M+1
            DO 10 I=1,M
              Q(I)=M•P(I)
              IF (Q(I) .LT. 1.0) THEN
                NN=NN+1
                W(NN)=I
              ELSE
                NP=NP−1
                W(NP)=I
              ENDIF
   10       CONTINUE
            DO 20 S=1,M−1
              I=W(S)
              J=W(NP)
              A(I)=J
              Q(J)=Q(J)+Q(I)−1.0
              IF (Q(J) .LT. 1.0) NP=NP+1
   20       CONTINUE
            A(W(M))=W(M)
            DO 30 I=1,M
```

```
30      Q(I)=Q(I)+I-1
        RETURN
        END

        FUNCTION ALRV (A,Q,M)
        INTEGER ALRV,A(M),M,I
        REAL Q(M),U
        U=M*RND( )
        I=1+INT(U)
        IF ( U .LE. Q(I)) THEN
          ALRV=I
        ELSE
          ALRV=A(I)
        ENDIF
        RETURN
        END
```

Index

Acceptance sampling, *see* Rejection sampling
ACT Sirius 1, 83, 215
Akaike's AIC Criterion, 156
Alias method, 72ff, 232f
Antithetic variates, 118, 124, 129ff, 139f
APL, 53, 83
Apple II, 18
Autoregressive spectral estimation, 156f

Bartlett's decomposition, 99, 117
BASIC, 5, 10, 17, 18, 76, 83, 97
Batching, 145, 150ff
Bays-Durham shuffle, 42, 51
BBC computer, 5, 10, 27, 76, 83, 97, 215
BCPL, 40
Beta distribution, 60f, 65, 90, 92f
Binomial distribution, 75, 78, 92
Blocks, 119, 138
Bootstrap, 4, 174ff
Box-Muller algorithm, 54, 93
Brownian motion, 107
Buffon's needle, 14, 193ff, 199

Cauchy distribution, 8, 60, 66f, 87, 92, 120
CDC, 83, 216
Chi-squared distribution, 55
Chi-squared test, 44, 95
Cholesky decomposition, 98
Closest pair, 6
Collective, 19
Combinatorial optimization, 179, 181
Combining generators, 43
Common random numbers, 119, 138
Composition methods, 63, 102
Conditional Monte Carlo, 136f
Conditioning, 118, 134ff

Confidence interval:
  bias-corrected percentile, 177
  Monte-Carlo, 176ff, 198
  percentile, 176
Congruential generator, 17
Continuous distributions, 81ff
Control variates, 118, 124ff
Correlation, extremal, 130
Correlation tests, 24, 45
Corvus Concept, 83, 215, 217
Coupon collector's test, 44
Cox process, 111
Cryptography, 16

Decimation, 27
Deletion, 143, 146
Discrepancies, 190
Discrete distributions, 71ff, 231ff
  multivariate, 100
Discrete-event simulation, 105, 169
Doubly stochastic Poisson process, 111
Dynamic thinning, 103

Eigenvalues, 188f, 199
Electronic noise, 15
Envelope, 61
Experimental design, 137ff
Exponential distribution, 55, 59, 63, 67, 69, 87, 230
Exponential spacings, 97

Factorial experiments, 139
Fast Fourier transform, 109
F distribution, 55, 90, 92
Fermat's little theorem, 47

Fibonacci recursion, 15, 51
Forsythe rejection method, 62, 65, 93
FORTRAN, 46, 78, 97, 100, 217

Gamma distribution, 88ff, 92, 95, 231
Gaps test, 43, 51
Gaussian processes, 105ff
Generalized feedback shift-register generators
    (GFSR), 28ff, 227ff
Geometric distribution, 77, 92
Gibbsian point process, 112
Gibbs sampler, 115
GLIM3, 40

Hazard function, 102
Heterogeneous Poisson process, 101
Hit-or-miss Monte Carlo, 121

IBM computers, 23, 38, 83
Image textures, 116
Importance sampling, 118, 122f, 139
Indexed search, 72
Infinitely divisible distributions, 91
Initial transient, 143, 146ff
Inverse Gaussian distribution, 94
Inversion method, 59, 71

Jacknife, 158, 160

k-distributed, 19, 29
Kolmogorov-Smirnov test, 44f, 92, 95

Languages for simulation, 10, 105
Lattice, 23, 33ff, 220
    polar, 37
Lifetime distributions, 102
Linear equations, 186ff

Markov process, 104
Markov random field, 114
Median, 7, 8, 98
Metropolis' method, 113
Middle-square method, 15f, 50
Minkowski's theorem, 36
Mod, 12, 217
Models, 2

Monte-Carlo:
    confidence intervals, 176ff, 198
    integration, 1, 119, 131
    tests, 4, 171ff, 198
Multivariate distributions, 98
    discrete, 100

NAG library, 38, 46
Neave effect, 55ff
Negative binomial distribution, 78
Normal distribution, 5, 54, 60, 65, 67, 82ff,
    92, 228ff
    contaminated, 9

Order statistics, 96ff

Pareto distribution, 104
PDP-11 computer, 23
Percentile confidence interval, 176
Period, 17
    full, 20
    maximal, 20
Permutations, 42, 81
Permutation test, 44, 51
PERT analysis, 134
PET computer, 18
Physical methods, 14f
Pincus' method, 180f, 198
Point pattern, 6, 26, 45
Point process, 110ff
Poisson distribution, 72, 76, 79, 92
Poisson process, 10, 12, 55, 100f, 107, 111
Polar lattice, 37
Polar method (for normals), 62
Power function, 124, 172
Pretesting, 69
Prewhitening, 157
Primality testing, 170
Primitive root, 21, 219
Princeton robustness study, 7, 13, 134
Pseudo-random, 2, 15

Quantiles, 98, 117
Quasi-Monte-Carlo integration, 189ff
Quasi-random, 15
Queueing systems, 10
    control and antithetic variates, 132ff
    with slow server, 161ff

Random fields, 113
Random numbers, 14
Random search algorithms, 179
Random sequences, 19
Random start algorithm, 180
RANDU, 23, 34, 41
Ratio of uniforms method, 65ff
Regenerative process, 159
Regenerative simulation, 12, 143, 157ff
Rejection sampling, 60ff
Renewal process, 100, 110, 140
Reporting simulation results, 4
Reservoir sampling, 80, 94
Response surface designs, 139, 185
Reusing uniforms, 93
Reversible Markov chain, 113
Rotation sampling, 141
Runs test, 44, 51

Sampling without replacement, 80f
Searching tables, 71
Seed, 20, 42
    matrix, 29
    switching, 140
Sequential tests, 122
Shift-register generator, 17, 26
Shuffling, 42
Simulated annealing, 181ff
Simulation, 1
    regenerative, 12, 143, 157ff
Slow server example, 161ff
Sorting methods, 98
Spectral density, 110, 144, 155ff, 168
Spectral test, 37
Spin exchange method, 116
Squeezing, 67ff
Stable distributions, 91. See also Cauchy
    distribution; Normal distribution

Standardized time series, 146, 153ff
Statistical experiment, 2, 4
Statistical inference, 4, 171ff
Steady-state simulation, 142
Stochastic algorithms, 170
Stochastic approximation, 185
Stochastic differential equations, 107
Stochastic optimization, 183ff
Stratified sampling, 118, 122, 131
"Student," 5
    t-distribution, 5, 8, 55, 87, 92, 94
Superposition, 102
Swindles, 119
Switching methods, 65
Sylvester's problem, 5

Table method (Marsaglia-Norman-Cannon),
    76f
Tausworthe generators, 27ff
t-distribution, see "Student"
Terminating approach to simulation, 142
Thinning method, 101
Time series:
    methods, 155ff
    models, 106
    standardized, 153f
Tours, 12, 135, 157
Turning band method, 108

Variance reduction, 4, 7, 9, 118, 197
VAX computer, 6, 38, 97, 163, 216, 217
Virtual measures, 135
Von Mises distribution, 60f
Von Neumann's method, 63, 87, 92, 230

Waiting-time paradox, 161
Weibull distribution, 60
Wishart distribution, 99

WILEY SERIES IN PROBABILITY AND STATISTICS

ESTABLISHED BY WALTER A. SHEWHART AND SAMUEL S. WILKS

Editors: *David J. Balding, Noel A. C. Cressie, Nicholas I. Fisher, Iain M. Johnstone, J. B. Kadane, Geert Molenberghs. Louise M. Ryan, David W. Scott, Adrian F. M. Smith, Jozef L. Teugels*
Editors Emeriti: *Vic Barnett, J. Stuart Hunter, David G. Kendall*

The *Wiley Series in Probability and Statistics* is well established and authoritative. It covers many topics of current research interest in both pure and applied statistics and probability theory. Written by leading statisticians and institutions, the titles span both state-of-the-art developments in the field and classical methods.

Reflecting the wide range of current research in statistics, the series encompasses applied, methodological and theoretical statistics, ranging from applications and new techniques made possible by advances in computerized practice to rigorous treatment of theoretical approaches.

This series provides essential and invaluable reading for all statisticians, whether in academia, industry, government, or research.

† ABRAHAM and LEDOLTER · Statistical Methods for Forecasting
  AGRESTI · Analysis of Ordinal Categorical Data
  AGRESTI · An Introduction to Categorical Data Analysis
  AGRESTI · Categorical Data Analysis, *Second Edition*
  ALTMAN, GILL, and McDONALD · Numerical Issues in Statistical Computing for the Social Scientist
  AMARATUNGA and CABRERA · Exploration and Analysis of DNA Microarray and Protein Array Data
  ANDĚL · Mathematics of Chance
  ANDERSON · An Introduction to Multivariate Statistical Analysis, *Third Edition*
* ANDERSON · The Statistical Analysis of Time Series
  ANDERSON, AUQUIER, HAUCK, OAKES, VANDAELE, and WEISBERG · Statistical Methods for Comparative Studies
  ANDERSON and LOYNES · The Teaching of Practical Statistics
  ARMITAGE and DAVID (editors) · Advances in Biometry
  ARNOLD, BALAKRISHNAN, and NAGARAJA · Records
* ARTHANARI and DODGE · Mathematical Programming in Statistics
* BAILEY · The Elements of Stochastic Processes with Applications to the Natural Sciences
  BALAKRISHNAN and KOUTRAS · Runs and Scans with Applications
  BARNETT · Comparative Statistical Inference, *Third Edition*
  BARNETT and LEWIS · Outliers in Statistical Data, *Third Edition*
  BARTOSZYNSKI and NIEWIADOMSKA-BUGAJ · Probability and Statistical Inference
  BASILEVSKY · Statistical Factor Analysis and Related Methods: Theory and Applications
  BASU and RIGDON · Statistical Methods for the Reliability of Repairable Systems
  BATES and WATTS · Nonlinear Regression Analysis and Its Applications
  BECHHOFER, SANTNER, and GOLDSMAN · Design and Analysis of Experiments for Statistical Selection, Screening, and Multiple Comparisons
  BELSLEY · Conditioning Diagnostics: Collinearity and Weak Data in Regression

*Now available in a lower priced paperback edition in the Wiley Classics Library.
†Now available in a lower priced paperback edition in the Wiley–Interscience Paperback Series.

† BELSLEY, KUH, and WELSCH · Regression Diagnostics: Identifying Influential Data and Sources of Collinearity

BENDAT and PIERSOL · Random Data: Analysis and Measurement Procedures, *Third Edition*

BERRY, CHALONER, and GEWEKE · Bayesian Analysis in Statistics and Econometrics: Essays in Honor of Arnold Zellner

BERNARDO and SMITH · Bayesian Theory

BHAT and MILLER · Elements of Applied Stochastic Processes, *Third Edition*

BHATTACHARYA and WAYMIRE · Stochastic Processes with Applications

† BIEMER, GROVES, LYBERG, MATHIOWETZ, and SUDMAN · Measurement Errors in Surveys

BILLINGSLEY · Convergence of Probability Measures, *Second Edition*

BILLINGSLEY · Probability and Measure, *Third Edition*

BIRKES and DODGE · Alternative Methods of Regression

BLISCHKE AND MURTHY (editors) · Case Studies in Reliability and Maintenance

BLISCHKE AND MURTHY · Reliability: Modeling, Prediction, and Optimization

BLOOMFIELD · Fourier Analysis of Time Series: An Introduction, *Second Edition*

BOLLEN · Structural Equations with Latent Variables

BOLLEN and CURRAN · Latent Curve Models: A Structural Equation Perspective

BOROVKOV · Ergodicity and Stability of Stochastic Processes

BOULEAU · Numerical Methods for Stochastic Processes

BOX · Bayesian Inference in Statistical Analysis

BOX · R. A. Fisher, the Life of a Scientist

BOX and DRAPER · Empirical Model-Building and Response Surfaces

* BOX and DRAPER · Evolutionary Operation: A Statistical Method for Process Improvement

BOX, HUNTER, and HUNTER · Statistics for Experimenters: Design, Innovation, and Discovery, *Second Editon*

BOX and LUCEÑO · Statistical Control by Monitoring and Feedback Adjustment

BRANDIMARTE · Numerical Methods in Finance: A MATLAB-Based Introduction

BROWN and HOLLANDER · Statistics: A Biomedical Introduction

BRUNNER, DOMHOF, and LANGER · Nonparametric Analysis of Longitudinal Data in Factorial Experiments

BUCKLEW · Large Deviation Techniques in Decision, Simulation, and Estimation

CAIROLI and DALANG · Sequential Stochastic Optimization

CASTILLO, HADI, BALAKRISHNAN, and SARABIA · Extreme Value and Related Models with Applications in Engineering and Science

CHAN · Time Series: Applications to Finance

CHARALAMBIDES · Combinatorial Methods in Discrete Distributions

CHATTERJEE and HADI · Sensitivity Analysis in Linear Regression

CHATTERJEE and PRICE · Regression Analysis by Example, *Third Edition*

CHERNICK · Bootstrap Methods: A Practitioner's Guide

CHERNICK and FRIIS · Introductory Biostatistics for the Health Sciences

CHILÈS and DELFINER · Geostatistics: Modeling Spatial Uncertainty

CHOW and LIU · Design and Analysis of Clinical Trials: Concepts and Methodologies, *Second Edition*

CLARKE and DISNEY · Probability and Random Processes: A First Course with Applications, *Second Edition*

* COCHRAN and COX · Experimental Designs, *Second Edition*

CONGDON · Applied Bayesian Modelling

CONGDON · Bayesian Statistical Modelling

CONOVER · Practical Nonparametric Statistics, *Third Edition*

COOK · Regression Graphics

---

*Now available in a lower priced paperback edition in the Wiley Classics Library.

†Now available in a lower priced paperback edition in the Wiley–Interscience Paperback Series.

COOK and WEISBERG · Applied Regression Including Computing and Graphics

COOK and WEISBERG · An Introduction to Regression Graphics

CORNELL · Experiments with Mixtures, Designs, Models, and the Analysis of Mixture Data, *Third Edition*

COVER and THOMAS · Elements of Information Theory

COX · A Handbook of Introductory Statistical Methods

* COX · Planning of Experiments

CRESSIE · Statistics for Spatial Data, *Revised Edition*

CSÖRGŐ and HORVÁTH · Limit Theorems in Change Point Analysis

DANIEL · Applications of Statistics to Industrial Experimentation

DANIEL · Biostatistics: A Foundation for Analysis in the Health Sciences, *Eighth Edition*

* DANIEL · Fitting Equations to Data: Computer Analysis of Multifactor Data, *Second Edition*

DASU and JOHNSON · Exploratory Data Mining and Data Cleaning

DAVID and NAGARAJA · Order Statistics, *Third Edition*

* DEGROOT, FIENBERG, and KADANE · Statistics and the Law

DEL CASTILLO · Statistical Process Adjustment for Quality Control

DeMARIS · Regression with Social Data: Modeling Continuous and Limited Response Variables

DEMIDENKO · Mixed Models: Theory and Applications

DENISON, HOLMES, MALLICK and SMITH · Bayesian Methods for Nonlinear Classification and Regression

DETTE and STUDDEN · The Theory of Canonical Moments with Applications in Statistics, Probability, and Analysis

DEY and MUKERJEE · Fractional Factorial Plans

DILLON and GOLDSTEIN · Multivariate Analysis: Methods and Applications

DODGE · Alternative Methods of Regression

* DODGE and ROMIG · Sampling Inspection Tables, *Second Edition*

* DOOB · Stochastic Processes

DOWDY, WEARDEN, and CHILKO · Statistics for Research, *Third Edition*

DRAPER and SMITH · Applied Regression Analysis, *Third Edition*

DRYDEN and MARDIA · Statistical Shape Analysis

DUDEWICZ and MISHRA · Modern Mathematical Statistics

DUNN and CLARK · Basic Statistics: A Primer for the Biomedical Sciences, *Third Edition*

DUPUIS and ELLIS · A Weak Convergence Approach to the Theory of Large Deviations

* ELANDT-JOHNSON and JOHNSON · Survival Models and Data Analysis

ENDERS · Applied Econometric Time Series

† ETHIER and KURTZ · Markov Processes: Characterization and Convergence

EVANS, HASTINGS, and PEACOCK · Statistical Distributions, *Third Edition*

FELLER · An Introduction to Probability Theory and Its Applications, Volume I, *Third Edition*, Revised; Volume II, *Second Edition*

FISHER and VAN BELLE · Biostatistics: A Methodology for the Health Sciences

FITZMAURICE, LAIRD, and WARE · Applied Longitudinal Analysis

* FLEISS · The Design and Analysis of Clinical Experiments

FLEISS · Statistical Methods for Rates and Proportions, *Third Edition*

† FLEMING and HARRINGTON · Counting Processes and Survival Analysis

FULLER · Introduction to Statistical Time Series, *Second Edition*

FULLER · Measurement Error Models

GALLANT · Nonlinear Statistical Models

GEISSER · Modes of Parametric Statistical Inference

GEWEKE · Contemporary Bayesian Econometrics and Statistics

---

*Now available in a lower priced paperback edition in the Wiley Classics Library.

†Now available in a lower priced paperback edition in the Wiley–Interscience Paperback Series.

GHOSH, MUKHOPADHYAY, and SEN · Sequential Estimation

GIESBRECHT and GUMPERTZ · Planning, Construction, and Statistical Analysis of Comparative Experiments

GIFI · Nonlinear Multivariate Analysis

GIVENS and HOETING · Computational Statistics

GLASSERMAN and YAO · Monotone Structure in Discrete-Event Systems

GNANADESIKAN · Methods for Statistical Data Analysis of Multivariate Observations, *Second Edition*

GOLDSTEIN and LEWIS · Assessment: Problems, Development, and Statistical Issues

GREENWOOD and NIKULIN · A Guide to Chi-Squared Testing

GROSS and HARRIS · Fundamentals of Queueing Theory, *Third Edition*

* HAHN and SHAPIRO · Statistical Models in Engineering

HAHN and MEEKER · Statistical Intervals: A Guide for Practitioners

HALD · A History of Probability and Statistics and their Applications Before 1750

HALD · A History of Mathematical Statistics from 1750 to 1930

† HAMPEL · Robust Statistics: The Approach Based on Influence Functions

HANNAN and DEISTLER · The Statistical Theory of Linear Systems

HEIBERGER · Computation for the Analysis of Designed Experiments

HEDAYAT and SINHA · Design and Inference in Finite Population Sampling

HELLER · MACSYMA for Statisticians

HINKELMANN and KEMPTHORNE · Design and Analysis of Experiments, Volume 1: Introduction to Experimental Design

HINKELMANN and KEMPTHORNE · Design and Analysis of Experiments, Volume 2: Advanced Experimental Design

HOAGLIN, MOSTELLER, and TUKEY · Exploratory Approach to Analysis of Variance

* HOAGLIN, MOSTELLER, and TUKEY · Exploring Data Tables, Trends and Shapes

* HOAGLIN, MOSTELLER, and TUKEY · Understanding Robust and Exploratory Data Analysis

HOCHBERG and TAMHANE · Multiple Comparison Procedures

HOCKING · Methods and Applications of Linear Models: Regression and the Analysis of Variance, *Second Edition*

HOEL · Introduction to Mathematical Statistics, *Fifth Edition*

HOGG and KLUGMAN · Loss Distributions

HOLLANDER and WOLFE · Nonparametric Statistical Methods, *Second Edition*

HOSMER and LEMESHOW · Applied Logistic Regression, *Second Edition*

HOSMER and LEMESHOW · Applied Survival Analysis: Regression Modeling of Time to Event Data

† HUBER · Robust Statistics

HUBERTY · Applied Discriminant Analysis

HUNT and KENNEDY · Financial Derivatives in Theory and Practice

HUSKOVA, BERAN, and DUPAC · Collected Works of Jaroslav Hajek— with Commentary

HUZURBAZAR · Flowgraph Models for Multistate Time-to-Event Data

IMAN and CONOVER · A Modern Approach to Statistics

† JACKSON · A User's Guide to Principle Components

JOHN · Statistical Methods in Engineering and Quality Assurance

JOHNSON · Multivariate Statistical Simulation

JOHNSON and BALAKRISHNAN · Advances in the Theory and Practice of Statistics: A Volume in Honor of Samuel Kotz

JOHNSON and BHATTACHARYYA · Statistics: Principles and Methods, *Fifth Edition*

JOHNSON and KOTZ · Distributions in Statistics

JOHNSON and KOTZ (editors) · Leading Personalities in Statistical Sciences: From the Seventeenth Century to the Present

JOHNSON, KOTZ, and BALAKRISHNAN · Continuous Univariate Distributions, Volume 1, *Second Edition*

JOHNSON, KOTZ, and BALAKRISHNAN · Continuous Univariate Distributions, Volume 2, *Second Edition*

JOHNSON, KOTZ, and BALAKRISHNAN · Discrete Multivariate Distributions

JOHNSON, KEMP, and KOTZ · Univariate Discrete Distributions, *Third Edition*

JUDGE, GRIFFITHS, HILL, LÜTKEPOHL, and LEE · The Theory and Practice of Econometrics, *Second Edition*

JUREČKOVÁ and SEN · Robust Statistical Procedures: Aymptotics and Interrelations

JUREK and MASON · Operator-Limit Distributions in Probability Theory

KADANE · Bayesian Methods and Ethics in a Clinical Trial Design

KADANE AND SCHUM · A Probabilistic Analysis of the Sacco and Vanzetti Evidence

KALBFLEISCH and PRENTICE · The Statistical Analysis of Failure Time Data, *Second Edition*

KASS and VOS · Geometrical Foundations of Asymptotic Inference

† KAUFMAN and ROUSSEEUW · Finding Groups in Data: An Introduction to Cluster Analysis

KEDEM and FOKIANOS · Regression Models for Time Series Analysis

KENDALL, BARDEN, CARNE, and LE · Shape and Shape Theory

KHURI · Advanced Calculus with Applications in Statistics, *Second Edition*

KHURI, MATHEW, and SINHA · Statistical Tests for Mixed Linear Models

* KISH · Statistical Design for Research

KLEIBER and KOTZ · Statistical Size Distributions in Economics and Actuarial Sciences

KLUGMAN, PANJER, and WILLMOT · Loss Models: From Data to Decisions, *Second Edition*

KLUGMAN, PANJER, and WILLMOT · Solutions Manual to Accompany Loss Models: From Data to Decisions, *Second Edition*

KOTZ, BALAKRISHNAN, and JOHNSON · Continuous Multivariate Distributions, Volume 1, *Second Edition*

KOTZ and JOHNSON (editors) · Encyclopedia of Statistical Sciences: Volumes 1 to 9 with Index

KOTZ and JOHNSON (editors) · Encyclopedia of Statistical Sciences: Supplement Volume

KOTZ, READ, and BANKS (editors) · Encyclopedia of Statistical Sciences: Update Volume 1

KOTZ, READ, and BANKS (editors) · Encyclopedia of Statistical Sciences: Update Volume 2

KOVALENKO, KUZNETZOV, and PEGG · Mathematical Theory of Reliability of Time-Dependent Systems with Practical Applications

LACHIN · Biostatistical Methods: The Assessment of Relative Risks

LAD · Operational Subjective Statistical Methods: A Mathematical, Philosophical, and Historical Introduction

LAMPERTI · Probability: A Survey of the Mathematical Theory, *Second Edition*

LANGE, RYAN, BILLARD, BRILLINGER, CONQUEST, and GREENHOUSE · Case Studies in Biometry

LARSON · Introduction to Probability Theory and Statistical Inference, *Third Edition*

LAWLESS · Statistical Models and Methods for Lifetime Data, *Second Edition*

LAWSON · Statistical Methods in Spatial Epidemiology

LE · Applied Categorical Data Analysis

LE · Applied Survival Analysis

LEE and WANG · Statistical Methods for Survival Data Analysis, *Third Edition*

LePAGE and BILLARD · Exploring the Limits of Bootstrap

*Now available in a lower priced paperback edition in the Wiley Classics Library.

†Now available in a lower priced paperback edition in the Wiley–Interscience Paperback Series.

LEYLAND and GOLDSTEIN (editors) · Multilevel Modelling of Health Statistics

LIAO · Statistical Group Comparison

LINDVALL · Lectures on the Coupling Method

LIN · Introductory Stochastic Analysis for Finance and Insurance

LINHART and ZUCCHINI · Model Selection

LITTLE and RUBIN · Statistical Analysis with Missing Data, *Second Edition*

LLOYD · The Statistical Analysis of Categorical Data

LOWEN and TEICH · Fractal-Based Point Processes

MAGNUS and NEUDECKER · Matrix Differential Calculus with Applications in Statistics and Econometrics, *Revised Edition*

MALLER and ZHOU · Survival Analysis with Long Term Survivors

MALLOWS · Design, Data, and Analysis by Some Friends of Cuthbert Daniel

MANN, SCHAFER, and SINGPURWALLA · Methods for Statistical Analysis of Reliability and Life Data

MANTON, WOODBURY, and TOLLEY · Statistical Applications Using Fuzzy Sets

MARCHETTE · Random Graphs for Statistical Pattern Recognition

MARDIA and JUPP · Directional Statistics

MASON, GUNST, and HESS · Statistical Design and Analysis of Experiments with Applications to Engineering and Science, *Second Edition*

McCULLOCH and SEARLE · Generalized, Linear, and Mixed Models

McFADDEN · Management of Data in Clinical Trials

* McLACHLAN · Discriminant Analysis and Statistical Pattern Recognition

McLACHLAN, DO, and AMBROISE · Analyzing Microarray Gene Expression Data

McLACHLAN and KRISHNAN · The EM Algorithm and Extensions

McLACHLAN and PEEL · Finite Mixture Models

McNEIL · Epidemiological Research Methods

MEEKER and ESCOBAR · Statistical Methods for Reliability Data

MEERSCHAERT and SCHEFFLER · Limit Distributions for Sums of Independent Random Vectors: Heavy Tails in Theory and Practice

MICKEY, DUNN, and CLARK · Applied Statistics: Analysis of Variance and Regression, *Third Edition*

* MILLER · Survival Analysis, *Second Edition*

MONTGOMERY, PECK, and VINING · Introduction to Linear Regression Analysis, *Third Edition*

MORGENTHALER and TUKEY · Configural Polysampling: A Route to Practical Robustness

MUIRHEAD · Aspects of Multivariate Statistical Theory

MULLER and STOYAN · Comparison Methods for Stochastic Models and Risks

MURRAY · X-STAT 2.0 Statistical Experimentation, Design Data Analysis, and Nonlinear Optimization

MURTHY, XIE, and JIANG · Weibull Models

MYERS and MONTGOMERY · Response Surface Methodology: Process and Product Optimization Using Designed Experiments, *Second Edition*

MYERS, MONTGOMERY, and VINING · Generalized Linear Models. With Applications in Engineering and the Sciences

† NELSON · Accelerated Testing, Statistical Models, Test Plans, and Data Analyses

† NELSON · Applied Life Data Analysis

NEWMAN · Biostatistical Methods in Epidemiology

OCHI · Applied Probability and Stochastic Processes in Engineering and Physical Sciences

OKABE, BOOTS, SUGIHARA, and CHIU · Spatial Tesselations: Concepts and Applications of Voronoi Diagrams, *Second Edition*

OLIVER and SMITH · Influence Diagrams, Belief Nets and Decision Analysis

PALTA · Quantitative Methods in Population Health: Extensions of Ordinary Regressions

---

*Now available in a lower priced paperback edition in the Wiley Classics Library.

†Now available in a lower priced paperback edition in the Wiley–Interscience Paperback Series.

PANKRATZ · Forecasting with Dynamic Regression Models

PANKRATZ · Forecasting with Univariate Box-Jenkins Models: Concepts and Cases

* PARZEN · Modern Probability Theory and Its Applications

PEÑA, TIAO, and TSAY · A Course in Time Series Analysis

PIANTADOSI · Clinical Trials: A Methodologic Perspective

PORT · Theoretical Probability for Applications

POURAHMADI · Foundations of Time Series Analysis and Prediction Theory

PRESS · Bayesian Statistics: Principles, Models, and Applications

PRESS · Subjective and Objective Bayesian Statistics, *Second Edition*

PRESS and TANUR · The Subjectivity of Scientists and the Bayesian Approach

PUKELSHEIM · Optimal Experimental Design

PURI, VILAPLANA, and WERTZ · New Perspectives in Theoretical and Applied
    Statistics

† PUTERMAN · Markov Decision Processes: Discrete Stochastic Dynamic Programming

QIU · Image Processing and Jump Regression Analysis

* RAO · Linear Statistical Inference and Its Applications, *Second Edition*

RAUSAND and HØYLAND · System Reliability Theory: Models, Statistical Methods,
    and Applications, *Second Edition*

RENCHER · Linear Models in Statistics

RENCHER · Methods of Multivariate Analysis, *Second Edition*

RENCHER · Multivariate Statistical Inference with Applications

* RIPLEY · Spatial Statistics

* RIPLEY · Stochastic Simulation

ROBINSON · Practical Strategies for Experimenting

ROHATGI and SALEH · An Introduction to Probability and Statistics, *Second Edition*

ROLSKI, SCHMIDLI, SCHMIDT, and TEUGELS · Stochastic Processes for Insurance
    and Finance

ROSENBERGER and LACHIN · Randomization in Clinical Trials: Theory and Practice

ROSS · Introduction to Probability and Statistics for Engineers and Scientists

† ROUSSEEUW and LEROY · Robust Regression and Outlier Detection

* RUBIN · Multiple Imputation for Nonresponse in Surveys

RUBINSTEIN · Simulation and the Monte Carlo Method

RUBINSTEIN and MELAMED · Modern Simulation and Modeling

RYAN · Modern Regression Methods

RYAN · Statistical Methods for Quality Improvement, *Second Edition*

SALEH · Theory of Preliminary Test and Stein-Type Estimation with Applications

* SCHEFFE · The Analysis of Variance

SCHIMEK · Smoothing and Regression: Approaches, Computation, and Application

SCHOTT · Matrix Analysis for Statistics, *Second Edition*

SCHOUTENS · Levy Processes in Finance: Pricing Financial Derivatives

SCHUSS · Theory and Applications of Stochastic Differential Equations

SCOTT · Multivariate Density Estimation: Theory, Practice, and Visualization

SEARLE · Linear Models for Unbalanced Data

SEARLE · Matrix Algebra Useful for Statistics

† SEARLE, CASELLA, and McCULLOCH · Variance Components

SEARLE and WILLETT · Matrix Algebra for Applied Economics

SEBER and LEE · Linear Regression Analysis, *Second Edition*

† SEBER · Multivariate Observations

† SEBER and WILD · Nonlinear Regression

SENNOTT · Stochastic Dynamic Programming and the Control of Queueing Systems

* SERFLING · Approximation Theorems of Mathematical Statistics

SHAFER and VOVK · Probability and Finance: It's Only a Game!

SILVAPULLE and SEN · Constrained Statistical Inference: Inequality, Order, and Shape
    Restrictions

---

*Now available in a lower priced paperback edition in the Wiley Classics Library.

†Now available in a lower priced paperback edition in the Wiley–Interscience Paperback Series.

SMALL and McLEISH · Hilbert Space Methods in Probability and Statistical Inference

SRIVASTAVA · Methods of Multivariate Statistics

STAPLETON · Linear Statistical Models

STAUDTE and SHEATHER · Robust Estimation and Testing

STOYAN, KENDALL, and MECKE · Stochastic Geometry and Its Applications, *Second Edition*

STOYAN and STOYAN · Fractals, Random Shapes and Point Fields: Methods of Geometrical Statistics

STYAN · The Collected Papers of T. W. Anderson: 1943–1985

SUTTON, ABRAMS, JONES, SHELDON, and SONG · Methods for Meta-Analysis in Medical Research

TAKEZAWA · Introduction to Nonparametric Regression

TANAKA · Time Series Analysis: Nonstationary and Noninvertible Distribution Theory

THOMPSON · Empirical Model Building

THOMPSON · Sampling, *Second Edition*

THOMPSON · Simulation: A Modeler's Approach

THOMPSON and SEBER · Adaptive Sampling

THOMPSON, WILLIAMS, and FINDLAY · Models for Investors in Real World Markets

TIAO, BISGAARD, HILL, PEÑA, and STIGLER (editors) · Box on Quality and Discovery: with Design, Control, and Robustness

TIERNEY · LISP-STAT: An Object-Oriented Environment for Statistical Computing and Dynamic Graphics

TSAY · Analysis of Financial Time Series, *Second Edition*

UPTON and FINGLETON · Spatial Data Analysis by Example, Volume II: Categorical and Directional Data

VAN BELLE · Statistical Rules of Thumb

VAN BELLE, FISHER, HEAGERTY, and LUMLEY · Biostatistics: A Methodology for the Health Sciences, *Second Edition*

VESTRUP · The Theory of Measures and Integration

VIDAKOVIC · Statistical Modeling by Wavelets

VINOD and REAGLE · Preparing for the Worst: Incorporating Downside Risk in Stock Market Investments

WALLER and GOTWAY · Applied Spatial Statistics for Public Health Data

WEERAHANDI · Generalized Inference in Repeated Measures: Exact Methods in MANOVA and Mixed Models

WEISBERG · Applied Linear Regression, *Third Edition*

WELSH · Aspects of Statistical Inference

WESTFALL and YOUNG · Resampling-Based Multiple Testing: Examples and Methods for p-Value Adjustment

WHITTAKER · Graphical Models in Applied Multivariate Statistics

WINKER · Optimization Heuristics in Economics: Applications of Threshold Accepting

WONNACOTT and WONNACOTT · Econometrics, *Second Edition*

WOODING · Planning Pharmaceutical Clinical Trials: Basic Statistical Principles

WOODWORTH · Biostatistics: A Bayesian Introduction

WOOLSON and CLARKE · Statistical Methods for the Analysis of Biomedical Data, *Second Edition*

WU and HAMADA · Experiments: Planning, Analysis, and Parameter Design Optimization

WU and ZHANG · Nonparametric Regression Methods for Longitudinal Data Analysis

YANG · The Construction Theory of Denumerable Markov Processes

* ZELLNER · An Introduction to Bayesian Inference in Econometrics

ZHOU, OBUCHOWSKI, and McCLISH · Statistical Methods in Diagnostic Medicine

*Now available in a lower priced paperback edition in the Wiley Classics Library.

†Now available in a lower priced paperback edition in the Wiley–Interscience Paperback Series.